Lecture Notes in Social Networks

More information about this series at http://www.springer.com/series/8768

Jim Duggan

System Dynamics Modeling with R

Springer

Jim Duggan
School of Engineering and Informatics
National University of Ireland Galway
Galway
Ireland

ISSN 2190-5428 ISSN 2190-5436 (electronic)
Lecture Notes in Social Networks
ISBN 978-3-319-34041-8 ISBN 978-3-319-34043-2 (eBook)
DOI 10.1007/978-3-319-34043-2

Library of Congress Control Number: 2016939926

Printed on acid-free paper

This Springer imprint is published by Springer Nature
The registered company is Springer International Publishing AG Switzerland

To Marie, Kate and James

Foreword

Since the emergence of system dynamics (SD) in the late 1950s, a range of literature has been published describing the methodology and detailing the best practices in model formulation together with its application to an ever-increasing span of domains. It would be true to say that the aspiring practitioner now has an enormous array of choices through which their competence in SD can be developed and broadened, far more so than faced those of us seeking to hone our skills in the 1970s and 1980s. Not only has the subject matter extended beyond the creation of formal simulation models to embrace the diagrammatic tools inherent in the qualitative aspects of the practice of SD (usually referred to as systems thinking) but also the simulation toolset on offer has similarly proliferated.

The student intending to become proficient at model formulation and execution is now faced with choices centered on the software platform to adopt. These extend from bespoke SD software to hybrid modeling tools which allow the user to code discrete-event and agent-based features in addition to SD. A software learning curve looms. Attempts to embrace SD modeling in, originally, general-purpose programming languages and, latterly, spreadsheets have not secured a significant user base.

In this new textbook, Jim Duggan breaks fresh ground in the practice of SD modeling by showing how it can be enabled through the *R* software environment for statistical computing and graphics. This software first emerged in the early 1990s, and it is chastening to realize that the scholarly endeavor inherent in this book could not have been mounted a mere twenty odd years ago. Being open source means of course that the *R* software is free, and thus, there exists significant potential to attract new students of SD as a consequence of this work. Not only that, but those whose predominant expertise is in the use of *R* for some other (data science) purpose could now find themselves being drawn into a whole new field as a result of this contribution.

The author's intent is clear: The book is devoted solely to the *formulation* of SD models. It is pitched at a technical level designed to showcase best practice in the craft of SD modeling with its underpinnings in integral calculus. Coverage firstly embraces the foundational aspects, followed by applications to various domains including economic growth, health care, and epidemiology. The final section is devoted to some technical aspects of SD modeling which embody mathematical and statistical methods and which are largely missing in other SD texts. Such aspects embrace model output analysis techniques, including statistical screening introduced by Ford and Flynn in 2005, as well as model calibration to reported data. These powerful and relatively new features of SD modeling are easily addressed using *R* and, of necessity, harness the strengths of its visualization features. Indeed, *R* is currently cited in the *System Dynamics Review* "Notes for Authors" as the recommended platform for producing publication quality graph plots.

The author is a current member of the Policy Council of the System Dynamics Society and serves on the editorial board of the *System Dynamics Review* as an Associate Editor. He has also acted as a Thread Chair for the methodology stream at system dynamics conferences. All this esteem stands as a testimony to his expertise in penning this welcome new perspective addressing the foremost component of SD practice—how to put together a set of model equations which reflect a dynamic system and which can then be used to explore its behavior over time. It is a valuable and significant addition to the SD literature.

Brian Dangerfield
Department of Management
University of Bristol, UK

Preface

A model should always be created for a purpose.

Jay W. Forrester, Urban Dynamics (1969), p. 113

System dynamics is a modeling approach used to construct simulation models of social systems, and these computerized models can then support policy analysis and decision making. This simulation method is based on calculus, and models of real-world dynamic processes are constructed using integral equations.

A key strength of system dynamics is that the simulation models provide an integrated view across organizational boundaries and functional areas, and so support a *joined-up thinking* approach to problem solving. System dynamics also provides a unique way of viewing social systems. This is known as the *feedback perspective,* where cause and effect between different system elements can be formally analyzed to help explain system behavior, and so generate insight into how to make better decisions. System dynamics has been successfully applied across a range of application areas, including complex and challenging domains such as project management, health care, manufacturing, epidemiology, and climate change.

The aim of this book was to provide readers with a practical understanding of system dynamics, so that they are in a position to design and implement simulation models in their chosen problem area. The book is structured into three thematic areas.

- *Foundations* (Chaps. 1–2). Chapter 1 provides an introduction to modeling and system dynamics. Foundational system dynamics concepts are presented, including simulation based on stocks, flows, and feedback. Models are solved using calculus, and the principles of numerical integration are presented. Chapter 2 is a primer in the open source R programming language and environment. R supports statistical computing and data analysis, and also has libraries for numerical integration. Important R concepts such as vectors, data frames, and functions are covered, and a system dynamics model is implemented in R.

- *Dynamic models of social systems* (Chaps. 3–5). Chapter 3 introduces a method for representing cause and effect equations in system dynamics. It then presents three different growth models in system dynamics, including s-shaped growth, an economic growth model, and a non-renewable resource growth and decline model. Chapter 4 introduces delays, which are features of social systems, and also the stock management heuristic for regulating important stock resources. A healthcare model combining three sectors, population, delivery, and general practitioner supply is specified, and this demonstrates how system dynamics can be applied to *joined-up* policy planning issues. Chapter 5 presents diffusion models for infectious disease transmission and control and includes the classic susceptible–infected–recovered (SIR) model. This is extended with a disaggregated model and highlights the power of R to simulate, using matrix manipulation, inter-cohort disease transmission dynamics.
- *Model testing and analysis* (Chaps. 6–7). Chapter 6 focuses on model testing and summarizes the system dynamics approach to validation. Practical methods for testing models are presented and implemented using R's unit test framework. Chapter 7 introduces a formal approach to feedback loop analysis. It presents a valuable parameter analysis method known as statistical screening, which uses a base set of sensitivity simulation runs to generate a data set that is analyzed using statistical methods. The results of this analysis then highlight those parameters that have the greatest influence on a variable's trajectory, which can enhance the overall policy design process and provide decision makers with more information on potential intervention strategies. This chapter also describes the important area of model calibration, where key parameters can be estimated in order to find the best fit of historical data to the underlying model structure.

System Dynamics and Calculus

System dynamics is grounded in calculus, which is the study of how things change over time. Calculus is described by Strogatz and Joffray (2009) as *perhaps the greatest idea that humanity has ever had*. Calculus allows us to communicate at the speed of light, build bridges across great divides, and take action to halt the spread of epidemics. Sterman (2000) observes that the study of calculus can be quite daunting, as the use of unfamiliar notation, and a focus on analytical solutions, can deter many people.

However, integration is an intuitive concept that can be understood without reference to formal mathematics, and system dynamics uses integration to model things that change over time. For example, system dynamics simulation models that generate projections for population levels in cities, prevalence values for infectious disease outbreaks, and inventory levels in global supply chains all use integration as the simulation method. In Chap. 1, the process of integration is summarized, with an initial look at *analytical* solutions, before focusing on *numerical* approaches, which are widely used in system dynamics simulation tools.

Related System Dynamics Texts

This book provides a complementary perspective to the range of system thinking and system dynamics textbooks, which include the work of Sterman (2000), Morecroft (2007), Warren (2008), Ford (1999), and Maani and Cavana (2007). This book's focus is on quantitative stock and flow models, and, similar to Meadows (2008), does not address the use of qualitative causal loop models. The motivation here is to focus on the set of core modeling concepts and constructs that can provide the necessary practical knowledge for readers to build system dynamics models.

Because of this, a number of areas covered by other texts, and by ongoing research in the *System Dynamics Review*[1] are not covered. These include model structures such as co-flows, bounded rationality, and supply line management; machine learning methods for analyzing system dynamics output, for example, techniques such as classification and clustering which can be used to explore the policy space (Kwakkel and Pruyt 2013); advanced analytical methods such as calibration, estimation, decision support, and optimization, which can support the model building process (Rahmandad et al. 2015); and formal model analysis using mathematical approaches such as eigenvalue and eigenvector analysis, which provide powerful formal methods to analyze the structure and behavior of system dynamics models (Duggan and Oliva 2013).

Related Complexity Work in Other Disciplines

In system dynamics, the definition of a complex system refers to a high-order, multiple-loop nonlinear feedback structure (Forrester 1969), and all social systems can be viewed from this perspective. The *order* is simply the number of stocks (or states) in the system, for example, Forrester's urban model is twentieth order. *Multiple-loop* reflects the presence of circular causal links between state variables, and the interaction among these loops can explain a complex system's behavior. It is important to acknowledge complementary computational methods for exploring and understanding complex systems. While system dynamics operates at an aggregate level that captures feedback, other methods, such as agent-based modeling, view a complex system from an individual perspective. Agents (e.g., individuals) are represented in a spatial network structure and make decisions based on

[1] http://onlinelibrary.wiley.com/journal/10.1002/(ISSN)1099-1727.

local information (Railsback and Grimm 2011). Epstein (2006) describes the classical agent-based experiment as follows:

> Situate an initial population of autonomous heterogeneous agents in a relevant spatial environment, allow them to interact according to simple local rules, and thereby generate—or "grow"—the macroscopic regularity from the bottom up.

This definition concisely summarizes the agent-based modeling perspective. By focusing on an autonomous agent (which is usually a model of a person or an organization), individual differences are captured and codified. For example, an agent-based model of infectious disease transmission would include a profile of different individuals (infants, young children, teenagers, adults, and elderly), their disease status (susceptible, infected, or recovered), a map of their contact network (family contacts, friendship links, and workplace connections), and a model of disease transmission based on the frequency of interactions between infected and susceptible people. From these interactions, an overall pattern of behavior emerges, and the outbreak of a disease can be traced, over time, through a causal chain of networked connections.

While a discussion of agent-based modeling is outside the scope of this text, there are parallels between system dynamics and agent-based modeling. Specifically, the disaggregated disease transmission model in Chap. 5, where the population is subdivided into age cohorts, has parallels with the agent-based perspective, and readers looking to bridge from system dynamics to agent-based modeling are encouraged to use the infectious disease case as an exemplar, and also consider other works that have explored similarities between the two methods, for example, the study by Rahmandad and Sterman (2008).

Why R?

Published system dynamics texts use the excellent set of available special-purpose modeling software to implement system dynamics models. In this text, an open source approach is used, and system dynamics models are implemented using R. R is a powerful programming language designed to analyze and interpret data, and it has an extensive set of open source libraries that can support decision analysis. This includes the **deSolve** library (Soetaert et al. 2010), which supports numerical integration using a range of numerical methods. There are three reasons for using R for system dynamics modeling:

- R provides a comprehensive set of statistical and optimization functions that can be used to analyze and calibrate simulation output. For example, in Chap. 7, the statistical screening method for system dynamics models (Ford and Flynn 2005) is implemented, as is a calibration method for data fitting. R also has a differential equation solver that can be used to implement system dynamics models.

- R has a powerful visualization library that can be used to present the behavior space of system dynamics models, and so present policy scenarios in a convincing manner to decision makers.
- R is a leading platform for data science methods such as regression and classification to support data analytics. By also supporting implementation of system dynamics models, it means that analysts can adopt multimethod approaches in addressing complex problems.

Model Catalog

One of the most enjoyable aspects of system dynamics modeling is that the method can be applied in a range of domains. Therefore, modelers are presented with opportunities to work across disciplines and interact with experts in a range of domains, on challenging policy problems. The models presented in this text illustrate the breadth of application of system dynamics and include the following:

- Epidemiology, with a focus on a contagious disease model in Chap. 5, and an interesting extension of this to a disaggregate form, based on a vectorized R implementation.
- Health systems design, which, in Chap. 4, provides a joined-up model comprising population demographics, a supply chain of general practitioners, and a demand-capacity model of general practitioner services to overall population.
- Economics and business, ranging from simple customer model in Chap. 1, and onto models of limits to growth, capital investment, and the impact of non-renewable resources on growth, all of which are covered in Chap. 3.

Intended Audience

This book can be used as a supporting text for courses in system dynamics, simulation, complexity, and mathematical modeling. Previous knowledge of basic calculus and an understanding of algebra would be an advantage, although in system dynamics, the stock and flow notation is intuitive and practical. The book also can be used as a reference for consultants and engineers who design and implement system dynamics models and plan to align their work with data science methods such as regression and classification. A full set of model and code examples, and lecture slides, is available online at https://github.com/JimDuggan.

Feedback

Comments, suggestions, and critiques are most welcome, including ideas for further examples that could be added to the online resource. Feedback can be emailed to jim.duggan@nuigalway.ie.

Acknowledgements

There are many individuals I would like to acknowledge who have contributed to my knowledge of systems thinking, system dynamics, and computer science. These include lecturers in Industrial Engineering in NUI Galway, who provided me with an early career insight into the decision support potential of management science, operations research, systems thinking, and simulation; colleagues in the College of Engineering and Informatics in particular, Gerry Lyons, Owen Molloy, and Enda Howley, for their excellent collaborations, and their shared enthusiasm for interdisciplinary research; and graduate research students for their innovation, ideas, and willingness to explore exciting research challenges at the intersection of system dynamics, data science, computer science, and complex social systems.

Thanks to my colleagues—from all parts of the world in the System Dynamics Society. The society provides a wonderful collegial space for sharing exciting ideas, investigating challenging research questions, and, of course, exploring simulation and modeling through stocks, flows, and feedbacks. In particular, thanks to Brian Dangerfield (University of Bristol), Pål Davidsen (University of Bergen), Bob Cavana (Victoria University of Wellington), and Rogelio Oliva (Texas A&M University) for their insights into system dynamics, their enthusiasm for the field, and their excellent advice on system dynamics research. Thanks also to the staff at Springer: Stephen Soehnlen, Senior Publishing Editor, for providing me with the opportunity to propose and write this book; and Pauline Lichtveld, Production Department, for her assistance in completing the production process. Finally, a special thank you to my family for their encouragement, inspiration, and support.

Galway, Ireland Jim Duggan
May 2016

References

Duggan J, Oliva R (2013) Methods for identifying structural dominance—introduction to the model analysis virtual issue. Syst Dyn Rev (Virtual Issue). http://onlinelibrary.wiley.com/journal/10.1002/(ISSN)1099-1727/homepage/VirtualIssuesPage.html

Epstein JM (2006) Generative social science: Studies in agent-based computational modeling. Princeton University Press. Chicago

Ford FA (1999) Modeling the environment: an introduction to system dynamics models of environmental systems. Island Press

Ford A, Flynn H (2005) Statistical screening of system dynamics models. Syst Dyn Rev 21:273–303

Forrester JW (1969) Urban dynamics. Pegasus Communications, Waltham, MA

Kwakkel JH, Pruyt E (2013) Exploratory modeling and analysis, an approach for model-based foresight under deep uncertainty. Technol Forecast Soc Change 80(3):419–431

Maani K, Cavana RY (2007) Systems thinking, system dynamics: managing change and complexity. Prentice Hall

Meadows DH (2008) Thinking in systems: a primer. Chelsea Green Publishing

Morecroft J (2007) Strategic modelling and business dynamics: a feedback systems approach. Wiley

Railsback SF, Grimm V (2011) Agent-based and individual-based modeling: a practical introduction. Princeton University Press

Rahmandad H, Oliva R, Osgood N (eds) (2015) Analytical methods for dynamic modelers. MIT Press, Cambridge

Rahmandad H, Sterman JD (2008) Heterogeneity and network structure in the dynamics of diffusion: comparing agent-based and differential equation models. Manag Sci 54(5):998–1014

Soetaert KER, Petzoldt T, Setzer RW (2010) Solving differential equations in R: package deSolve. J Stat Softw 33

Sterman JD (2000) Business dynamics: systems thinking and modeling for a complex world. Irwin/McGraw-Hill, Boston

Strogatz S, Joffray D (2009) The calculus of friendship: what a teacher and a student learned about life while corresponding about math. Princeton University Press

Thompson KM, Tebbens RJD (2008) Using system dynamics to develop policies that matter: global management of poliomyelitis and beyond. Syst Dyn Rev 24(4):433–449

Warren K (2008) Strategic management dynamics. Wiley, Chicago

Contents

Chapter 1
An Introduction to System Dynamics

Everything we do as individuals, as an industry, or as a society is done in the context of an information-feedback system.
Jay W. Forrester, Industrial Dynamics (1961), p. 15.

Abstract This chapter presents important concepts underlying the system dynamics modeling method. Following an initial definition of the term model, a summary of a successful system dynamics intervention is described. The key elements of system dynamics—*stocks and flows*—are explained. The process for simulating stock and flow models—*integral calculus*—is described, with an example of a company's customer base used to illustrate how stocks change, through their flows, over time. A summary of dimensional analysis for stock and flow equations is provided before the second feature of system dynamics modeling—*feedback*—is presented. The chapter concludes by summarizing the system dynamics methodology, which is a five-stage iterative process that guides model design, development, test and policy design.

Keywords Models · Stocks · Flows · Feedback · Integration

Models

Pidd (1996, p. 15) defines a model as:

> an external and explicit representation of part of reality as seen by the people who wish to use that model to understand, to change, to manage and to control that part of reality.

This is an insightful definition that also applies to system dynamics. The model building process focuses on a *part of reality* that needs to be understood and managed, and creates an *external and explicit representation,* in the form of a model, of this reality. This reality could be an organization faced with declining market share, a public health agency confronted by an infectious disease outbreak, or governments challenged by increased levels of carbon in the atmosphere, with

J. Duggan, *System Dynamics Modeling with R*, Lecture Notes in Social Networks,
DOI 10.1007/978-3-319-34043-2_1

the resulting rise in mean global temperatures. In these scenarios, decision makers are faced with a complex, and highly interacting, social system. Models provide a basis for decision makers to understand their world as an interconnected system, and to test out the impact of policy interventions *in silico*. Understanding leads to insight, and an opportunity to change, manage and control the system of interest.

In order for a model to be useful to decision makers, it must provide some view on future behavior, and Meadows et al. (1974) provide a valuable classification of the types of outputs models can provide:

- *Absolute, precise predictions*, for example, when and where will the next solar eclipse be observable?
- *Conditional, precise predictions,* for example, if a cooling systems fails in a nuclear power plant, what will be the maximum pressure exerted on the reactor's containment vessel?
- *Conditional, imprecise projections of dynamic behavior*, for example, if an infectious disease spreads through a population, what is the likely future burden of demand on intensive care facilities one month from the outbreak date?

Because system dynamics is primarily a technique for business and policy simulation modeling (Homer 2012), its primary focus is on the third class of model: those simulation models that provide conditional, imprecise projections of dynamic behavior. This is because social and business systems are by their nature unpredictable in the absolute sense (Meadows et al. 1974). So while *all models are wrong* (Box 1976), as they cannot generate precise point-predictions of future events in social systems, the challenge is to create *models that are useful* through extensive testing, benchmarking against available data, and continual iteration between experiments with the virtual world of simulation and the real world (Sterman 2002). System dynamics has a rich tradition of creating useful models across many disciplines, and, to illustrate this, an application of system dynamics to public health policy is presented.

System Dynamics in Action: Population Health Policy

In their paper—*Using system dynamics to develop policies that matter: global management of poliomyelitis and beyond*—Thompson and Tebbens (2008) document their award-winning research which demonstrates how system dynamics impacted global health policy analysis. This supported the Global Polio Eradication Initiative (GPEI) to eradicate wild polioviruses, which aimed to replicate the success of the eradication of smallpox in 1979 (Breman and Arita 1980). Initial results of this initiative, based on an intensive vaccination campaign, led to a reduction from 350,000 global annual cases to 1000 cases per year.

However, the eradication project faced funding shortfalls during 2002–3, and the allocation of vaccination resources prioritized endemic countries where the virus circulated, leaving other countries vulnerable. This containment policy inevitably led to further outbreaks. While additional investment was made to regain lost ground, a new policy debate started that questioned the feasibility of eradication, and suggested that the guiding policy should switch to one of control, as this could save resources while maintaining outbreak cases at manageable low levels.

The authors proceeded to evaluate the impact of this proposed policy change, and assess the economic impact, and potential disease burden, of these two distinct policy options. Core to the analysis was a system dynamics disease outbreak model. This model represented the population as a set of stocks and flows, where people were classified as being susceptible to, infected with or recovered from the wild poliovirus. The stock and flow model distinguished between 25 different age groups. This model was also informed by their prior studies related to risk management, including a cost-based analyses of tradeoffs associated with outbreak response.

As a result of the model building process, the authors highlighted the impact of *wavering*. This describes a scenario whereby in the context of successful vaccinations comes the perception that a high level of continued investment in vaccine administration is not required. This view was represented in the system dynamics model. Focusing on two Northern Indian states, two policy options were evaluated. The first was to vaccinate extensively until disease eradication, the second was to vaccinate only if the costs per incidence remained outside an acceptable level. The simulation demonstrated the impact of these two policies, and showed that the containment option leads to more cases, and costs, over a 20-year time horizon. Therefore, the model provided evidence to support the policy of eradication.

The next stage of the process involved the authors presenting their model and results at a stakeholder consultation, convened by the WHO Director-General Dr. Margaret Chan. The goal of this meeting was to consider the option of switching from eradication to control. Their author's system dynamics model, with its quantitative approach, long time horizon, and its analysis of the impact of wavering commitment, supported the case to continue the eradication policy, and this subsequently led to further resources to implement the eradication policy.

There are three modeling insights from this case study. First, it shows how system dynamics can be successfully applied to real-world problems, and achieve an impact in terms of policy analysis and implementation. Second, it demonstrates the importance of model purpose, where the modeling activity was focused on the core issue of whether to eradicate or control a disease. Third, the paper provides an excellent sense of the interdisciplinary skills required to build system dynamics models. The authors invested considerable time to understand the specific problem, and were well-positioned to defend their work to national and international policymakers, financial donors, fellow modelers, economists and epidemiologists. Furthermore, reflecting on Pidd's (1996) earlier definition of a model, it is evident that their system dynamics model has the following characteristics:

- It was an *external and explicit representation* of a problem, in that the model was a system dynamics representation of infectious disease transmission, based on the classic Susceptible-Infected-Recovered (SIR) model, which is covered in detail in Chap. 5.
- It focused on the *part of reality* that was important to the *people who wished to use the model* (key stakeholders), namely how best to improve overall population health in vulnerable countries through implementing the most appropriate disease containment policy.
- It provided a basis *to understand, change, manage and control* that part of reality by demonstrating the detrimental impact of wavering on future outbreaks, and so provided evidence to support the continuation of the eradication policy.

In summary, this highlights the potential of system dynamics modeling to make a difference to society. Within the field there are many documented examples of how this simulation method has been successfully applied across a range of disciplines, including business systems, project management, energy policy and health care. A further advantage of the system dynamics method is that it is grounded in the theory of dynamic systems, and in particular, it uses calculus—and the ideas of stocks and flows—to generate quantitative projections of a system's behavior over time.

Stocks and Flows

A stock is the foundation of any system (Meadows 2008), and stocks and flows are the building blocks of system dynamics models. They characterize the state of the system under study, as well as providing the information upon which decisions and actions are based (Sterman 2000). Stocks can only change through their flows, which are the quantities added to (inflow), or subtracted from (outflow), a stock over time. Stocks are present in many business and social systems, and examples include:

- Warehouse inventory (stock keeping units), which is the amount of stock within the four walls of a warehouse at a given point in time. Inflows include inventory arriving from suppliers (stock keeping units/week), and returns sent back by customers (stock keeping units/week). Outflows are goods shipped to customers (stock keeping units/week).
- Employees in an organization (people) across all processes and functions. Inflows include new hires (employees/month). Examples of outflows from this stock include employee retirements (people/month), employee attrition (people/month) and employee redundancies (people/month).
- The number of people suffering from an illness in the population. Inflows are the number of people becoming ill during a time period (people/week). Outflows are the number of people that recover over each time period (people/week).

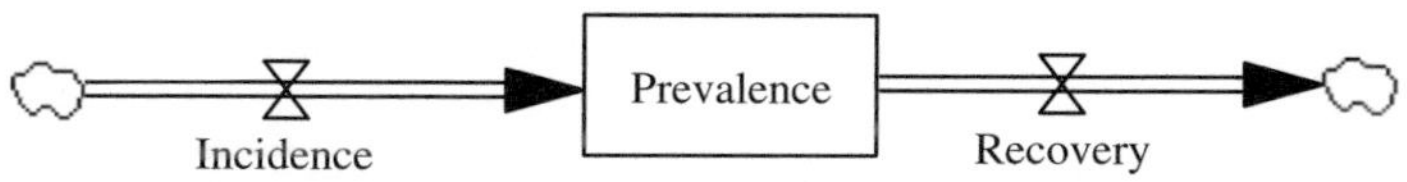

Fig. 1.1 A stock and flow model of illness in a population

Note that in all cases the units of the flows are the *units of the stock* divided by the *time period*. This time period is determined by the system under study, and can vary from seconds to years, depending on the problem's time horizon. In order to explore the stock and flow concept, an example from public health is presented, where the focus is on the presence of illness in a population. The stock and flow model for this is shown in Fig. 1.1.

The model visualization shows the stock as a *containers*, and the flows as *pipes* filling and draining this container, where the flow rates are controlled by valves. The variable names used for this initial model are informed by public health professionals, and it is usually good practice to build a model that practitioners can identify with. Therefore the following definitions are used (Giesecke 1994).

- *Prevalence* is defined as the number of people who have that disease at a specific time, and this is a stock. For example if the model captured the dynamics of seasonal influenza, this would be the number of people infected with influenza.
- *Incidence* is defined as the number of people who become ill with a certain disease during a defined time period. For seasonal influenza, this is usually measured each week, and the units are therefore (people/week). Incidence is a flow.
- *Recovery* is the number of people removed from the ill population per time period. Recovery is a flow, and its units are (people/week).

A feature of this one-stock model is that it can be used to highlight three principles of stock and flow systems. These ideas relate the behavior of the stock to the values of the net flow, where the net flow is the difference between all inflows and all outflows. For example, in a given week, if 1000 people contact influenza and 800 people recover from their bout of the virus, the net flow for the week is +200, which is the difference of the two flows. Because the difference is greater than zero, the prevalence will rise over this time period. Therefore, in the general case of any stock and flow system, the following conditions hold true:

- When the total sum of all inflows to a stock *is greater than* the total sum of all outflows, the stock will rise.
- When the total sum of all inflows to a stock *is less than* the total sum of all outflows, the stock will fall.
- When the total sum of inflows to a stock *equals* the total sum of outflows, the stock will remain unchanged. This is an interesting and often desired state of many systems, and is known as *dynamic equilibrium*.

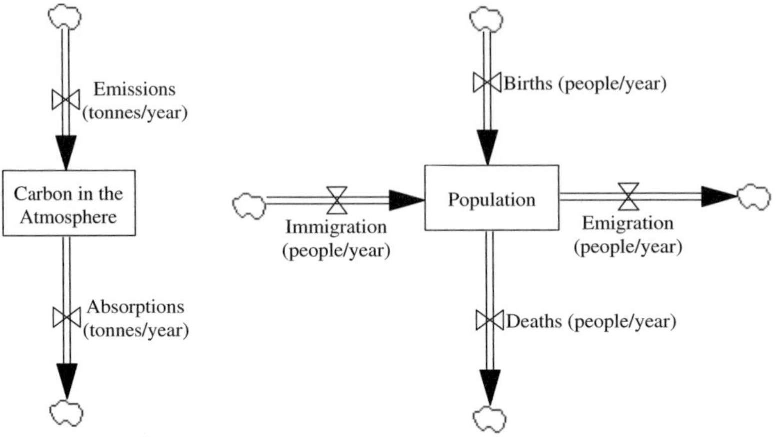

Fig. 1.2 Further examples of stock and flow systems

These principles can be applied to any system that changes over time, including challenges related to global warming, economics, and population planning. For example, Fig. 1.2 shows a model of carbon in the earth's atmosphere (a stock). This stock is increased by emissions, and reduced by absorptions. As the earth's carbon absorption rate is currently less than the carbon emissions rate, the amount of carbon in the atmosphere is increasing, and this is now shown to impact global temperatures. The second model describes, at a highest level of aggregation, the population of a country, with inflows of births and immigration, and outflows of deaths and emigration.

These two models are high-level, and represent the system of interest by a single stock. However, stock models can also be disaggregated to reveal finer-grained dynamics. Disaggregation is an important part of system dynamics modeling, and is necessary when there are sufficient differences in subsets of a variable, for example, cohorts in a population. This is shown in Fig. 1.3, where a country's population is broken down into age cohorts, and the stocks are cascaded in order to capture the dynamics of aging. Disaggregated population model structures such as this are particularly useful when exploring long-term dynamics of health systems, where age is an important determinant of health. To simplify the model, migrations are excluded, with the main focus on how the age profile of the population changes over time.

While these stock and flow models may appear straightforward—which is beneficial from a model building viewpoint—an important challenge is to formulate the inflows and outflows. For example the following questions must be addressed:

- How are delays in a system modeled, where items stay in a stock for a period of time and then progress?
- How are rate variables such as the number of births modeled, particularly when variables may depend on other stocks in the system?

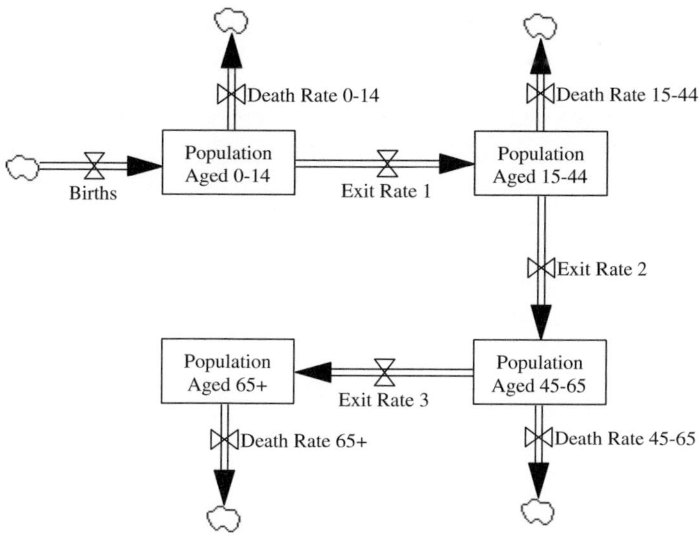

Fig. 1.3 A disaggregated model of a population, excluding migration

- How are decisions in a system modeled, where a manager decides, for example, how many new hires to take on in order to replenish the employee stock, and so maintain a company's resource base, and capacity to deliver services to customers?

Flow structures such as *fractional increase* and *fractional decrease* are explored later in this chapter, and Chap. 4 will describe formulating delays, and how to model management decisions such as stock replenishment.

In summary, stocks are present in many social systems. They represent accumulations, and can only change through their inflows and outflows. Stocks are solved using the mathematical process known as *integration*, and this is how system dynamics models are simulated.

Integration

Integration is the mathematical process of calculating the area under the net flow curve, between initial and final times. There are two main methods for integrating. The first method is analytical, where an integral is expressed as an equation that can be used to determine the stock's value at any future point in time. The second approach is numerical, which is commonly used for more complex higher-order (i.e. many stocks) systems, and is the method that will be used throughout this text.

The two methods are now explored, using a linear net flow equation, visualized in Fig. 1.4, where $f(t) = 2t$. Therefore the net flow starts at 0, and climbs to 20 after 10 time units. A quick visual inspection, using the formula for calculating the area of a triangle, will show that the integral after 20 time units is 0.5 * 20 * 10 = 100.

In order to solve this analytically, the standard integration method can be used (1.1). To achieve this, the net flow is represented as a derivative (1.2), with a corresponding indefinite integral solution (1.3) is found through applying (1.1). However, in this case the time interval is known, and therefore the area between two specific points can be evaluated as the difference in the indefinite integral solution over the time interval, and this is shown to be 100 in (1.4).

$$\int t^n dt = \frac{1}{n+1} t^{n+1} + c \tag{1.1}$$

$$\frac{\mathrm{dy}}{\mathrm{dt}} = 2\mathrm{t} \tag{1.2}$$

$$y = \int 2t\, dt = t^2 + c \tag{1.3}$$

$$y_{t=10} = |_0^{10}\, t^2 = 10^2 - 0^2 = 100 \tag{1.4}$$

The analytical solution shown in (1.4) can be used to calculate the stock's value at any future time interval. However, as already discussed, exact analytical solutions may not be feasible for higher-order, non-linear stock and flow systems. Approximate solutions can be calculated, and a widely-used numerical algorithm is known as *Euler's method.*

Euler's approach estimates the area under the net flow curve through a sequence of rectangles of identical width. The rectangle height is *the opening value of the net flow applied over the interval DT*, where DT is also known as the time step. As the time step gets smaller, the overall numerical solution becomes more accurate. Euler's equation accumulates the successive areas of these rectangles (1.5) by assuming that the net flow is constant over each time interval (the opening value of the net flow is taken).

$$Stock_t = Stock_{t-dt} + (Inflow_{t-dt} - Outflow_{t-dt}) * DT \tag{1.5}$$

Figure 1.4 uses a time step of 1 (normally this would be too large a value to use for an accurate simulation). From the time series plot, the sequence of successive rectangles is shown, and the stock's value is simply the summation of these rectangle areas, based on (1.5). The solution process is summarized in Table 1.1, which also shows the error term (the difference between the approximate integration and the true integration). In this example, the error term is the sum of the small triangle areas between the blue and red lines. This error term can be reduced by selecting a smaller time step, usually for system dynamics simulations a time step value of 1/8 or 1/16 is used.

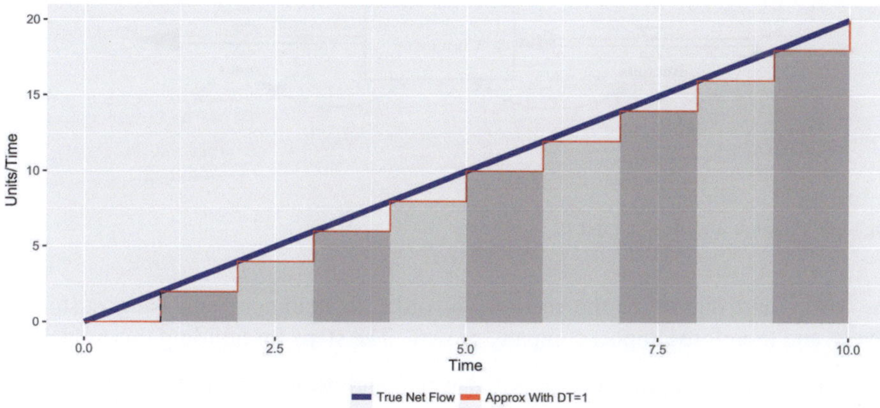

Fig. 1.4 Representation of an integration problem

Table 1.1 Approximate values of the integral of dy/dt = 2t, dt = 1, Euler's method

Time	$Stock_t$	Net flow	$Stock_A = t^2$	Error
0	0	0	0	0
1	0 + 0 = 0	2	1	1
2	0 + 2 = 2	4	4	2
3	2 + 4 = 6	6	9	3
4	6 + 6 = 12	8	16	4
5	12 + 8 = 20	10	25	5
6	20 + 10 = 30	12	36	6
7	30 + 12 = 42	14	49	7
8	42 + 14 = 56	16	64	8
9	56 + 16 = 72	18	81	9
10	72 + 18 = 90	20	100	10

In summary, integration is the basis for all system dynamics simulation runs. Once a model is expressed in terms of stocks and flows, the integration process is applied to every stock, for each time step. Therefore when all the initial stock values are known, and each flow has a defined equation, the integration process will simulate the behavior of all model variables.

A System Dynamics Model of Customers

In order to demonstrate how a system dynamics model is constructed, a one-stock model of an organization's customer base is modeled. Given that the customer base is an accumulation, it can be modeled as a stock. The inflow is recruits, and the outflow are losses, also known as the churn rate. The goal of organizations is to limit

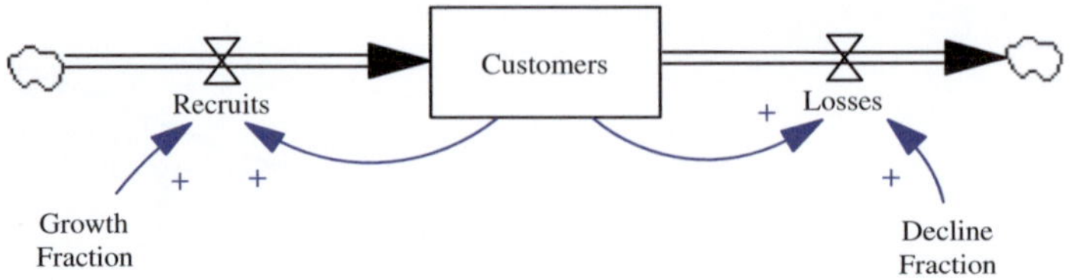

Fig. 1.5 A stock and flow model of customers

the losses and maximize the recruits, in order to maintain increasing customers levels, and therefore support company growth. The steps for building this model are:

- Identify the stock, provide an initial value, and decide on the flows that change the stock
- Formulate equations for the flows
- Decide on the time units, for example, is the simulation in days, months or years.
- Decide on the time interval, which is the start and finish time of the simulation run.

The stock and flow model is shown in Fig. 1.5, and the information dependencies between equations are shown, along with the type of relationship. For example, the "+" sign at the end of a link indicates that the variables move in the same direction. These type of causal links will be described shortly, and are important when considering the feedback structures of system dynamics models. The stock is expressed as an integral function, where the arguments are the inflows less the outflows, followed by the initial value. In effect, equation (1.6) is the similar to that shown earlier in (1.5). Stock equations are usually the most straightforward to formulate, as they can only change via their flows. The initial value of the stock for the simulation run is required, otherwise the integration process could not proceed.

$$Customers = INTEGRAL(Recruits - Losses, 10000) \tag{1.6}$$

Following the stock definition, all that remains is to formulate the inflow and outflow, and any auxiliary variables that they may depend on. An auxiliary variable is one that is not a stock or a flow, and is generally used to simplify flow equations. For most modelers, the most challenging task in system dynamics is the composition of flow and auxiliary equations (Dangerfield 2014). Conveniently, there are a number of pre-defined flow equation structures that can be used. In this case, two ideas will be used to formulate the inflow and outflow (Sterman 2000). These are:

- The *fractional increase rate*, where the inflow to a stock is proportional to the stock.
- The *fractional decrease rate*, where the outflow of a stock is proportional to the stock.

For the customer model, these can be viewed as reasonable assumptions. For example, all companies have annual expansion goals, where they seek to increase their customer base by a target growth fraction. On the other hand, companies are faced with the challenge of retaining customers, and therefore will seek to minimize the churn rate, or the fraction of customers that are lost each year. The flow equations can be formulated to reflect this real-world scenario. The inflow (1.7) is the product of the customers and the growth fraction, and this is a commonly used structure in system dynamics models.

$$Recruits = Customers \times Growth\,Fraction \tag{1.7}$$

The multiplier of the inflow is the growth fraction (1.8), and, for this example, this value varies over time, through the use of the *STEP* function. The *STEP* function, which is available in all system dynamics software, has the form: *STEP* (*<amount>*,*<time>*), and changes a variables value by <amount> at the specified simulation time <time>. In this case, the growth fraction starts at 0.07, drops to 0.03 at 2020, and drops by a further 1 % to 0.02 in 2025. In a more complex model, this growth fraction could depend on other system variables, for example, the number of marketing resources, product quality, and the size of the potential market. When an auxiliary does not directly depend on another model variable it is termed an *exogenous variable*. This type of variable will be discussed in greater detail later in this chapter.

$$Growth\,Fraction = 0.07 - STEP(0.04{,}2020) - STEP(0.01{,}2025) \tag{1.8}$$

The losses are formulated as a fixed proportion of the customer stock, and this is shown in (1.9). The decline fraction is fixed at 3 %, and this is captured in (1.10).

$$Losses = Customers \times Decline\,Fraction \tag{1.9}$$

$$Decline\,Fraction = 0.03 \tag{1.10}$$

This finalizes the model formulation, with five equations for the simulation model. The equations are complete, as all the variables shown in Fig. 1.5 are specified. There are no gaps, no ambiguities, just concrete equations that will simulate the customer model. All that remains is to decide on the simulation run settings, which are the time interval (2015–2030), the time step DT (0.25), and the time units (years). The model can then be simulated using a number of approaches, and in this case R's **deSolve** library was used. The simulation output is shown in Fig. 1.6.

It is worth reflecting on the simulation output in terms of how the stock behaves over time, which can be classified into three different phases.

- *Phase 1*, from 2015–2020, where the stock increases, as the net growth fraction is $0.07 - 0.03 = 0.04$. While this growth may look linear, it is in fact exponential, similar to how compound interest is calculated for a bank savings

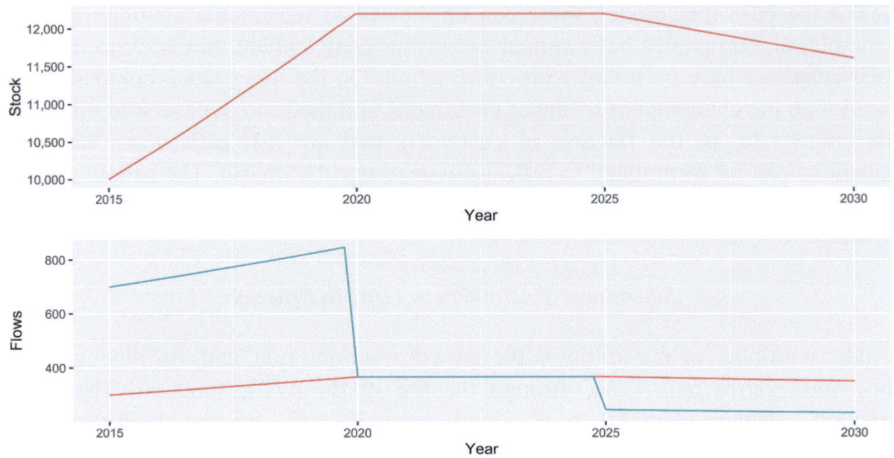

Fig. 1.6 Simulation output from the customer model

account. For example, it can be shown that solving the differential equation $dy/dx = gY$, where g is the fractional increase rate, yields the resulting integral equation solution $Y_t = Y_0 e^{gt}$, which confirms that in the stock growth is exponential.

- *Phase 2,* from 2020–2025, where the stock remains constant (dynamic equilibrium), given that the growth and decline fractions are equal, and cancel one another out. In this case the model is in dynamic equilibrium.
- *Phase 3,* from 2025–2030, where the decline fraction exceeds the growth fraction, and this results in a declining stock over time, as the net flow is negative. This decline in the stock is exponential, as it can be shown that solving the differential equation $dy/dx = -rY$, where r is the fractional decrease rate, yields the resulting integral equation solution $Y_t = Y_0 e^{-rt}$, where confirms that the stock decline follows an exponential decay pattern.

What is noteworthy about the three phases is that they confirm the fundamentals of stock and flow systems. If the inflow exceeds the outflow (i.e. time interval 2015–2020), the stock rises; if the inflow equals the outflow (i.e. time interval 2020–2025), the stock remains in equilibrium; and, if the outflow exceeds the inflow (time interval 2025–2030), the stock falls. While this is a simple model, these concepts are relevant to any system dynamics model, and can be applied to more complex models to support policy analysis and design. For example, the comparison of inflows to outflows forms part of epidemic threshold calculations, and this will be presented in Chap. 5.

Dimensional Analysis for Stock and Flow Equations

In the physical sciences and engineering, any equation representing a real-world process needs to have the units (i.e. dimensions) balanced on each side of the '=' sign (Dangerfield 2014). This checking—also known as dimensional analysis—is also an important activity in system dynamics, as it provides an excellent validation mechanism for any simulation model. As a starting point, the units for system stocks are identified. For example, stock units from a range of modeling challenges in business, health and the environment are shown in Table 1.2.

Stocks change through their flows, and therefore, in order to maintain dimensional consistency, a flow must have units of the stock it feeds, *divided by the units in which time is measured* (Coyle 1996). The selection of time unit depends on the problem being explored, for example, planning in a higher education context has annual student intake, therefore the most suitable time unit would be *year*. However, measuring the spread of an infectious disease such as influenza is typically performed on a weekly basis. Societal challenges such as global warming and efforts to controlling the amount of carbon in the atmosphere can have a time horizon measured in decades, and even centuries. An indication of flows, and their units is provided in Table 1.3, covering a wide range of applications areas.

Once the units for stocks and flows are identified, dimensional analysis can be performed, where both sides of an equation are simplified to their basic units. If the two sides of the dimensional equation are equal, then the equation is *dimensionally*

Table 1.2 Sample stock variables along with indicative values for units

Application area	Stock	Units
Business	Inventory	Stock keeping unit (SKU)
Financial planning	Cash	€, $
Education planning	Students	People
Epidemiology	Infected	People
Demographics	Population	People
Climate change	Carbon in the atmosphere	Metric tons

Table 1.3 Sample flow variables along with indicative values for units

Stock	Inflow	Outflow	Flow units
Inventory	Arrivals	Shipments	SKU/week
Cash	Deposits	Withdrawals	€/day, $/day
Student	Registrations	Graduations	People/year
Infected	Incidence	Recovery	People/day
Population	Births	Deaths	People/year
Carbon in the atmosphere	Emissions	Absorptions	Metric Tons/year

consistent. To illustrate the idea, the customer model from Fig. 1.5 is used, and the integral equation, similar to the format shown in (1.5), is shown in (1.11).

$$\begin{aligned} \mathit{Customers}_t &= \mathit{Customers}_{t-dt} + (\mathit{Recruits} - \mathit{Losses}) * \mathit{DT} \\ \text{People} &= \text{people} + (\text{people/year} - \text{people/year}) * \text{year} \end{aligned} \tag{1.11}$$

This equation is dimensionally consistent, as the inflow and outflow denominator (year) cancels with the dimensions of DT (year) to arrive at the dimension (people). This process also applies to flows in system dynamics models. Once the units of the flow multiplied by the time units equal the stock units, the stock equations will be dimensionally consistent. However, it is not sufficient just to have stock equations dimensionally accurate, *all model variables* should have their units checked and validated. For example, the equation for *recruits* (1.7) and *losses* (1.9) also need to be checked for dimensional consistency.

$$\begin{aligned} \mathit{Recruits} &= \mathit{Customers} * \mathit{Growth\,Fraction} \\ \text{people/year} &= \text{people} * (\text{people/year})/\text{person} \end{aligned} \tag{1.12}$$

$$\begin{aligned} \mathit{Losses} &= \mathit{Population} * \mathit{Decline\,Fraction} \\ \text{people/year} &= \text{people} * (\text{people/year})/\text{person} \end{aligned} \tag{1.13}$$

In (1.12) and (1.13), *recruits* and *losses* are flows, and therefore their respective units are (people/year). The units of the growth and decline fractions are (1/year), as these values are based on the number of people added/removed each year, divided by the number there to start with, which yields dimensions of (person/year)/person, or (1/year) (Dangerfield 2014). Therefore, the two flow equations are dimensionally consistent, and the customer model passes its dimensionality test. Software packages for system dynamics support dimensional checking, so adding in units at an early stage can improve the model building process. Later in Chap. 6, additional methods for validating system dynamics models are explored, where the benefit is to improve the model quality, and enhance client confidence. In the next section, the second foundational concept of system dynamics is summarized. This idea provides valuable insight to guide decision making in complex systems, and is known as *feedback*.

Feedback

Feedback is a defining element of system dynamics (Lane 2006), and identifying feedback loops in social systems is an important part of modeling building. Meadows (2008) describes a feedback loop as:

> A closed chain of causal connections from a stock, through a set of decisions or rules or physical laws or actions that are dependent on the level of the stock, and back again through a flow to change the stock.

A feedback loop is a chain of circular causal links, where the level of a stock influences a flow, which in turn will change the stock. The stock can influence the flow directly, or that influence could be determined through a series of intermediate auxiliary variables. Feedback processes are present in many systems. Earlier, when discussing stocks and flows, a warehouse example was presented. This can be examined in more detail to uncover a feedback process in operation.

In the warehouse, there is a quantity of products on shelves, and this quantity can be modeled as a stock. The company would have a target quantity of product to store, to ensure that stockouts would not happen, and to maintain high levels of customer satisfaction. For example, this target value could be two weeks of expected demand. At regular intervals (perhaps once per week), the warehouse manager would note the current level of the stock, and compare this to the target value. If more stock was needed, orders would be made from suppliers. These orders would then arrive at the warehouse, and their arrival would be modeled as an inflow to the stock. This inflow increases the stock, and so completes the feedback process that connects the stock to the inflow.

Consider a home heating system, and how its feedback process operates (Fig. 1.7). The occupant sets the desired room temperature. A heat sensor records the actual room temperature, and this is relayed to a controller. The controller logic determines if the temperature is lower that the desired. If it is the heater is activated, and the generated heat raises the room temperature. As the room temperature rises, the sensor detects monitors towards the desired value, and once this value is reached, the heater is switched off.

This is a further example of feedback, where the level of a stock (heat in the room) determines the amount of heat added (the flow) which in turn changes the heat in the room (the stock). It is an example of a *goal seeking system,* in that once a target is established the system is continually moved towards that target. These are known as *negative feedback loops,* and are annotated using the balancing (B) icon on the stock and flow model.

Loop polarity can be evaluated for any feedback loop, by examining the individual links contained in that loop. A link captures a cause and effect relationship

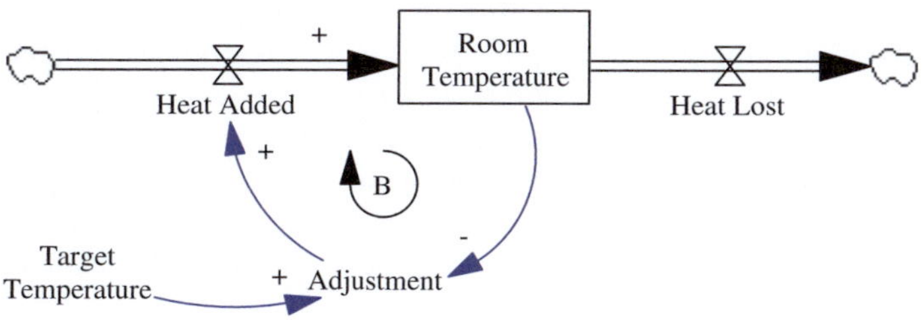

Fig. 1.7 Temperature control: an example of negative feedback

Fig. 1.8 Causal links and their polarity

Table 1.4 Tracing the changes of variables through the room temperature loop

Room temperature	↓	Adjustment	↑
Adjustment	↑	Heat added	↑
Heat added	↑	**Room temperature**	↑

between two variables (e.g. x and y), and an individual link can be either positive or negative. A positive link occurs when, all else being equal, the cause x increases, the effect y increases above what it would have been. A negative link means that as the cause x increases, then the effect y decreases below what it would have been (Sterman 2000). In the room temperature model, the feedback loop contains positive and negative links. A positive link occurs when the cause and effect move in the *same direction*, for example, as the adjustment increases, so to does the amount of heat added. A negative link implies that the cause and effect move in *opposite directions*, for instance, as the temperature rises, the adjustment falls (Fig. 1.7).

Calculating loop polarity is a straightforward task. The loop is broken down into a set of the causal links, and the impact of a change in one variable is traced through the causal chain, and back to the original variable. In this example, the loop contains three variables: *Room Temperature*, *Adjustment*, and *Heat Added.*

Table 1.4 shows the impact of a change in room temperature, where a variable in the loop can either rise (↑) or fall (↓). Assuming that the room temperature is falling due to heat loss (due to the stock outflow), the impact of this change through the feedback loop is as follows:

- As the *room temperature* decreases, the *adjustment* (which is the difference between desired and actual temperature) increases, as it is a negative link because the two values move in opposite directions.
- With an increase in *adjustment*, the amount of *heat added* also increases, as this is a positive link where the cause and effect move in the same direction.
- An increase in *heat added* then leads to an increase in *room temperature,* as this is also a positive link.

The individual link polarities combine to determine the overall loop polarity. With one iteration through the loop, the direction of the original variable has been impacted. In this case, at the outset the room temperature was falling, and following the sequence of circular causal links, the temperature rises. Room temperature has moved in the *opposite direction* after one iteration through the loop. This is an example of a regulating system, or more generally, *negative feedback.* A negative feedback loop also has an odd number of negative links (in this case 1), and this heuristic can be used to quickly calculate loop polarity.

The loop polarity calculation can be applied to a different model (Fig. 1.9), involving the interplay between capital and output, often termed the engine of economic growth (Meadows 2008).

The more machines and factories (capital) there are, the more goods and services (output) that can be produced. This model contains a set of circular causal links, as the loop contains three variables: *Capital*, *Output*, and *Investment in Capital.*

Table 1.5 traces the behavior of the loop's variables, from an initial starting point where we assume that the capital is increasing. The causal links are follows:

- As the *capital* increases, so to does *output,* and this is a positive link, as the variables move in the same direction.
- With an increase in *output*, the inflow *investment in capital* will increase. Again, this is a positive link.
- As *investment in capital* increases, the amount of capital (i.e. the stock) also increases, and this final link in the feedback loop is also positive.

In contrast to the room temperature example, the direction of change of capital has been *reinforced* or amplified as a consequence of the loop. Increased capital, through a cycle of reinvestment, leads to more capital. This is a classic example of *positive feedback*, which drives exponential growth, and terms such as *virtuous cycle* and *success to the successful* are often used (where the effect is desirable). On the other hand, positive feedback can also have detrimental effects (e.g. a run on a bank), where a value spirals out of control, and in this case the term *vicious cycle* is used instead. A positive feedback loop will always have an even (including zero) number of negative links, and this can be a useful shortcut taken in order to calculate loop polarity.

In summary, a complex system is an interlocking structure of feedback loops, and this loop structure is found many real-world processes (Forrester 1969). In particular:

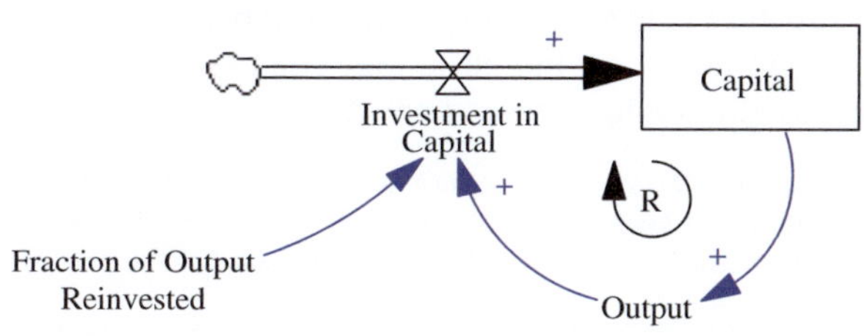

Fig. 1.9 Capital growth: an example of positive feedback

Table 1.5 Tracing the changes of variables through the capital investment loop

Capital	↑	Output	↑
Output	↑	Investment in capital	↑
Investment in capital	↑	**Capital**	↑

- A feedback loop is a closed chain of causal links from a stock, through a flow, and back to the original stock again.
- There are two classes of feedback loops. Negative feedback counteracts the direction of change, whereas positive feedback amplifies change and drives exponential growth.
- Loop polarity is calculated by evaluating the individual link polarities in a circular causal chain. If there are an odd number of negative links, the loop polarity in negative, otherwise the loop polarity is positive.

Modeling Feedback

Creating feedback models in system dynamics is challenging. It requires domain knowledge, and the skill to see the interrelationships between different system elements. The goal is to identify those feedbacks that influence overall system behavior. Forrester (1968) defines an important principle, centered on the idea of a system boundary:

> In concept a feedback system is a closed system. Its dynamic behavior arises within its internal structure. Any interaction which is essential to the behavior mode must be included inside the system boundary.

This definition provides a valuable context for identifying feedback structures. The challenge is to extend an initial stock and flow model, by adding additional stocks, flows and feedbacks that influence the systems behavior. This is done through collaborating with multiple stakeholders, who provide different perspectives and knowledge on the problem being addressed. As part of this interaction, people share their understanding of what variables need to be included within the model boundary. At some point in the model building process, there should be consensus that all the relevant stocks, flows and feedbacks have been included within the model boundary. This structure is then the closed system representation of the problem.

Richardson (2011) builds on this idea, and emphasizes the importance of the system boundary by discussing what is known as the *endogenous point of view*:

> The most salient aspect of the system dynamics approach are undoubtedly stocks and flows and feedback loops. These visible elements stand out and demand our attention. But it is worth noting that feedback loops are really a consequence of the endogenous point of view.

Endogenous refers to the idea that actions are caused by factors from *inside of the system*. With the endogenous viewpoint behavior can be explained through the system's feedback structure, and not through the actions of an external, uncontrollable, exogenous source. Sterman (2002) writes that system dynamics practitioners are trained to be suspicious of exogenous variables, and they must challenge model constants in order to see whether they could be part of the feedback structure. This process of *challenging the constants* is central to the endogenous perspective, and can be used to discover important feedback loops.

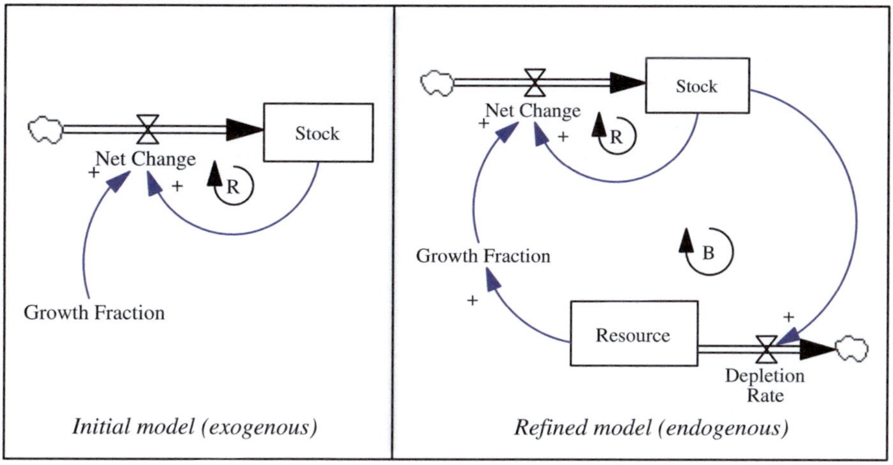

Fig. 1.10 Evolving an endogenous feedback perspective

In order to provide an example of how the endogenous point of view can be used to identify feedback structures, a one-stock model is presented, as shown on the left in Fig. 1.10. This stock increases based on the growth fraction, and is structurally similar to the capital growth model shown earlier in Fig. 1.9. The growth fraction is a constant, and is exogenous, as the value has its source *outside of the system.* In other words, this exogenous variable is not influenced by any other model variable. With a constant growth rate, the system stock will grow exponentially, with no physical limits. However, growth without limits is unrealistic, as for any system, there are always factors that limit growth. Therefore, the flawed assumption of this initial model is that the growth fraction never changes. By taking the endogenous perspective, this assumption can be challenged, and a new version of the model generated.

The model boundary is expanded to include other stocks that may impact the systems behavior. The target of enquiry now becomes the constant (exogenous) variable *growth fraction*. In this case, the following question can be asked: *what is the growth fraction dependent on?* In this generic model, it is assumed that the growth fraction depends on the availability of a non-renewable resource. There are well-documented cases, such as the population growth and decline on Easter Island (Brandt and Merico 2015), where stocks have grown based on the availability of non-renewable resources, only to decline once those resources were consumed. From this we can extend the model in three ways:

- The growth fraction depends on the resource availability, where resources are a stock. This is a positive link. More resources lead to a higher growth rate.
- The resource depletion rate depends on the level of the stock. This is positive link. The higher the stock, the greater the depletion rate.

- The resource is reduced by the depletion rate. This is a negative link, as a higher depletion rate leads to a reduction in stock. In this model, the resource is assumed to be non-renewable, as there is no inflow to replenish lost resources.

This is an example of extending the model boundary, has revealed a new, and significant, feedback structure. What was previously an exogenous variable (growth fraction) is now endogenous. As a result, there is now a more realistic model that links, via feedback, two system stocks that are clearly interdependent. Based on this endogenous feedback model, we can also determine the polarity of the new feedback loop by taking a variable of interest and tracing the impact of its increase through each feedback loop.

The first feedback loop in summarized in Table 1.6. As the stock's change is reinforced after a single iteration, this is a positive feedback loop, and so will drive exponential growth or decline. The second feedback loop, which emerged as a result of focusing on the exogenous variable *growth fraction,* is summarized in Table 1.7. This shows that the direction of change for the variable of interest (Stock) is reversed following a cycle through the loop. Therefore, this is a negative feedback loop that acts as a limiting factor to the stocks growth.

This example highlights the process for expanding model boundaries, which can then ensure that important feedbacks are considered throughout the modeling process. The limits to growth model is explored in further detail in Chap. 3, where a stock grows rapidly based on a resource, but as the resource diminishes, the stock enters a period of rapid decline. Furthermore, a healthcare model is formulated in Chap. 4, and feedbacks identified between the different model sectors.

It is also worth reiterating that modeling feedback in system dynamics is challenging, and the interested reader is recommended to follow up with excellent examples of feedback thinking from the system dynamics literature. These include:

- How system dynamics models can help the public policy process (Ghaffarzadegan et al. 2011).
- Identifying feedback structures in the project management process (Lyneis and Ford 2007), and,
- System dynamics models applied to understand population health outcomes (Homer 1993).

Table 1.6 Calculating the polarity for first feedback loop

Stock	↑	Net change	↑
Net change	↑	Stock	↑

Table 1.7 Calculating the polarity for second feedback loop

Stock	↑	Depletion rate	↑
Depletion rate	↑	Resource	↓
Resource	↓	Growth fraction	↓
Growth fraction	↓	Net change	↓
Net change	↓	Stock	↓

The Model Building Process

The starting point for system dynamics is that the model must be created for a specific purpose (Forrester 1969). For example, consider the following three scenarios:

- In the food industry, a multinational company might want to improve its supply chain performance, and a system dynamics model could provide ways to evaluate different production and distribution strategies.
- In the public health domain, stakeholders may wish to assess the impact of mass vaccination to protect citizens against an outbreak of a virulent influenza virus.
- In the airline industry, a company could face a challenge of declining passenger numbers, and look for ways to assess the impact of further capacity investment.

The common thread across the these examples is that each model addresses a clear problem, and therefore each model has a definite purpose. With a clearly defined goal, a valuable strength of system dynamics is that it is supported by an iterative five-stage methodology. This can be used to manage projects and ensure a process for engaging with clients and their problems owners in order to address their problems.

The model building process comprises five inter-related activities, shown in Fig. 1.11, and based on Morecroft (2007).

1. *Articulate Problem.* This involves identifying the specific problem with clients, and exploring the reasons why it is a problem worth addressing. Important variables are selected, and the appropriate time horizon identified. Historical data is gathered to support this initial analysis, and behavior over time graphs can provide valuable input into this problem articulation stage.

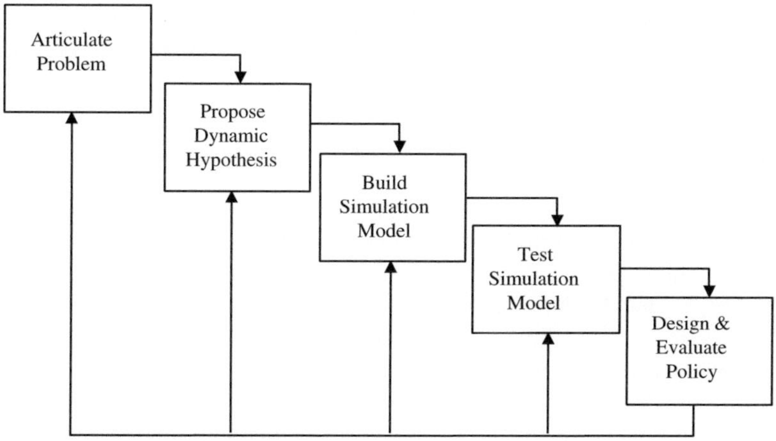

Fig. 1.11 The system dynamics modeling process

2. *Propose Dynamic Hypothesis.* Following on from this initial stage, a *dynamic hypothesis* is proposed where the aim is to identify the stock, flow and feedback structures can best explain the problematic behavior. The problem is mapped using tools such as causal loop diagrams, stock and flow maps, and other appropriate facilitation tools.
3. *Build Simulation Model.* With a mapping structure and feedbacks identified, the *simulation model* can be formulated, with the stock and flow structure and decision rules. Tasks such as parameter estimation (covered in Chap. 7) and initial tests can be performed, and feedback gained from clients.
4. *Test Simulation Model.* The fourth stage is *testing,* where the model behavior is compared to the known reference models, and its robustness is tested under extreme conditions. Chapter 6 shows how extreme condition tests can be designed and implemented. Sensitivity testing can also be used to evaluate the impact of uncertainly in model parameters on overall outcomes.
5. *Design and Evaluate Policy.* The fifth stage is *policy design and evaluation,* which requires that the model is robust and has passed a suite of rigorous tests. In this activity, new decision rules, strategies and structures that could be implemented in the real-world are evaluated. The simulation model is used to perform *what-if analysis* to observe the potential impact of new policies. Following this, improvement actions can be agreed with clients, and the implementation of system changes can follow.

While the process flow would indicate a linear sequence from problem definition to policy design, the reality is that building system dynamics models is a highly iterative process. Smaller, high-level, models may be initially mapped and implemented, and the revised based on feedback from stakeholders. The idea of having a final unchanging model is unrealistic, as models are always in a continuous state of evolution, were “each question, each reaction, each new input of information, and each difficulty in explaining the model leads to modification, clarification, and extension” (Forrester 1985). Further insights into the model building process is provided by Vennix (1996), who offers excellent guidance on how to perform group model building, and so manage a system dynamics intervention with multiple stakeholders.

Summary

This chapter provided an introduction to system dynamics. This simulation method is based on finding stocks, flows and feedbacks that are relevant to the problem of interest. The technical solution process used is integration, where stocks accumulate their inflows, less any outflows. The process of finding feedback by exploring the system boundary was introduced, as was the overall five-stage problem solving process. System dynamics equations can be solved using special purpose simulation tools. In this text the R framework is used to solve equations, and an introduction to R is presented in Chap. 2.

Exercises

1. The net flow for a population is given $dP/dt = rP$, where r is the fractional growth rate. From this, show that the integral is given by $P_t = P_0e^{rt}$ where P_0 is the initial value of the population.
2. Create a two stock system for a University. One stock models students, the other staff. Identify inflows and outflows for each stock. Add an additional variable to the model called *student staff ratio*. Higher values of this ratio make the University less attractive for students, and also result in the University hiring more staff. Show any feedback loops, and calculate the loop polarities using two methods.
3. Consider the net flow $dy/dt = 4t$. Assuming the stock y is initially zero, solve analytically for the value of y after 10 time units. Use Euler's equation, with DT = 0.5, to solve for y.

References

Box GE (1976) Science and statistics. J Am Stat Assoc 71(356):791–799

Brandt G, Merico A (2015) The slow demise of Easter Island: insights from a modeling investigation. Front Ecol Evol 3:13

Breman JG, Arita I (1980) The confirmation and maintenance of smallpox eradication. N Engl J Med 303(22):1263–1273

Coyle RG (1996) System dynamics modelling: a practical approach. CRC Press, Boca Raton

Dangerfield B (2014) Systems thinking and system dynamics: a primer. In: Discrete-event simulation and system dynamics for management decision making. Wiley, New York City, pp 26–51

Forrester JW (1961) Industrial dynamics. MIT Press, Cambridge (Reprinted by Pegasus Communications: Waltham, MA)

Forrester JW (1968) Principles of systems. Pegasus Communications, Waltham

Forrester JW (1969) Urban dynamics. Pegasus Communications, Waltham

Forrester JW (1985) The "model versus a modeling" process. Syst Dyn Rev 1(1):133–134

Ghaffarzadegan N, Lyneis J, Richardson GP (2011) How small system dynamics models can help the public policy process. Syst Dyn Rev 27(1):22–44

Giesecke J (1994) Modern infectious disease epidemiology. Edward Arnold (Publisher) Ltd., London

Homer JB (1993) A system dynamics model of national cocaine prevalence. Syst Dyn Rev 9(1):49–78

Homer J (2012) Models that matter: selected writings on system dynamics, 1985–2010. Grapeseed Press, New York

Lane DC (2006) IFORS' operational research hall of Fame Jay Wright Forrester. Int Trans Oper Res 13(5):483–492

Lyneis JM, Ford DN (2007) System dynamics applied to project management: a survey, assessment, and directions for future research. Syst Dyn Rev 23(2–3):157–189

Meadows DH (2008) Thinking in systems: a primer. Chelsea Green Publishing, White River Junction, Vermont

Meadows DL, Behrens WW, Meadows DH, Naill RF, Randers J, Zahn E (1974) Dynamics of growth in a finite world. Wright-Allen Press, Cambridge

Morecroft J (2007) Strategic modelling and business dynamics: a feedback systems approach. Wiley, New York

Pidd M (1996) Tools for thinking: modelling in management sciences. Wiley, New York

Richardson GP (2011) Reflections on the foundations of system dynamics. Syst Dyn Rev 27(3):219–243

Sterman JD (2000) Business dynamics: systems thinking and modeling for a complex world. Irwin/McGraw-Hill, Boston

Sterman JD (2002) All models are wrong: reflections on becoming a systems scientist. Syst Dyn Rev 18(4):501–531

Thompson KM, Tebbens RJD (2008) Using system dynamics to develop policies that matter: global management of poliomyelitis and beyond. Syst Dyn Rev 24(4):433–449

Vennix JA (1996) Group model building: facilitating team learning using system dynamics, vol 2001. Wiley, Chichester

Chapter 2
An Introduction to R

> *Exploration is our mission; we and those who use our software want to find new paths to understand the data and the underlying processes.*
> John M. Chambers, Software for Data Analysis (2008, p. 3).

Abstract This chapter introduces R, a dialect of the S language, which was developed at Bell Laboratories. R's inventor Dr. John Chambers was awarded the 1998 Association of Computing Machinery Software award. In its citation, the ACM noted that *S will forever alter the way people analyze, visualize, and manipulate data*. R's *mission* is to enable the best and most thorough exploration of data possible. R is open-source software (GNU General Public License), and has statistical, data manipulation, and visualization libraries. R is a functional programming language, where software programs are organized into *functions* that can be invoked to transform data. This chapter describes key R elements, including vectors, lists, matrices, data frames and functions. It concludes by presenting a system dynamics model of customer growth, which is implemented using the **deSolve** open source package. Appendix A summarizes the installation process for R, and the reader is recommended to work through this chapter using the R Studio console, so that the short examples can be executed.

Keywords Vectors · Functions · Matrices · Data frames · deSolve

Vectors

The fundamental data type in R is the vector, which is a variable that contains a sequence of elements that have the same data type (Matloff 2009). A vector is defined by the ability to index its elements by position, in order to extract or replace a subset of data (Chamber 2008). The vector object is similar to a one-dimensional array structure in a programming language such as C or Java. Vectors can be created in the following manner.

J. Duggan, *System Dynamics Modeling with R*, Lecture Notes in Social Networks,
DOI 10.1007/978-3-319-34043-2_2

```
v1<-c(1,2,3,4,5)
```

This creates a vector variable **v1** and assigns it an initial value using the function **c**, which is the combine function in R. By typing **v1** at the console, the vector's values can be inspected.

```
> v1
[1] 1 2 3 4 5
```

The printed value **[1]** at the beginning of the output is a useful piece of information that displays the starting index for that particular printed row of vector data. The concept of an index is important in R, as it allows access to individual elements of a vector, using the square brackets notation. In R the index for a vector starts at 1. This command displays the third element of the vector **v1**.

```
> v1[3]
[1] 3
```

In R, variable types can include integer, numeric, character, and logical types. The mode of a variable can be examined using the **typeof(x)** function call. In a vector, the mode of each element is the same.

```
> typeof(v1)
[1] "double"
```

Functions can operate on vectors, for example, to find the length, maximum value, and minimum value, the functions **length(x), max(x)** and **min(x)** are used. Each of these R functions returns a vector of length 1.

```
> length(v1)
[1] 5

> max(v1)
[1] 5

> min(v1)
[1] 1
```

A powerful feature of R is that it supports *vectorization*, where functions can operate on every element of a vector, and return the results of each individual operation in a new vector. Many in-built R functions support vectorization, including the square root function **sqrt(x).**

```
> v1
[1] 1 2 3 4 5

> r<-sqrt(v1)

> r
[1] 1.000000 1.414214 1.732051 2.000000 2.236068
```

A significant benefit of this feature is that the analyst does not have to write a loop to iterate through the vector. Vectorized functions have the general form of *vector in, vector out* (Matloff 2009), where the size of the output vector mirrors the size of the input vector.

Arithmetic operations can also be applied to vectors in an element-wise manner. For this example, the vector **v1** is multiplied by the constant 3, and the result (**v2**) is then added to **v1,** and finally stored in **v3**.

```
> v1
[1] 1 2 3 4 5
> v2<-3*v1
> v2
[1]  3  6  9 12 15
> v3<-v1+v2
> v3
[1]  4  8 12 16 20
```

When operations are applied to two vectors that requires them to be of equal length, R automatically recycles the shorter vector until it is of sufficient length to match the longer one.

```
> v4<-c(10,20)
> v1
[1] 1 2 3 4 5
> v5<-v1+v4
Warning message:
In v1 + v4 :
  longer object length is not a multiple of shorter object
length
> v5
[1] 11 22 13 24 15
```

Conditional expressions can also be applied to vectors, and these are used to filtering vector data. For example, by taking the original vector **v1** and applying a conditional expression to that vector, R will return a logical vector (e.g. a vector whose elements are either TRUE or FALSE) containing the results for each conditional expression evaluation. In this case, the condition tests which vector elements are even, and R's modulus operator (%%) is used.

```
> v1
[1] 1 2 3 4 5
> test<-v1 %% 2 == 0
> test
[1] FALSE  TRUE FALSE  TRUE FALSE
```

An interesting feature of R is that this logical vector can now be used as an *index* to the original vector, and those values that match to TRUE in the logical vector will be returned by the operation. Using the NOT logical operator (!), all the FALSE values can be returned.

```
> evens<-v1[test]
> evens
[1] 2 4

> odd<-v1[!test]
> odd
[1] 1 3 5
```

As R is a functional programming language, many operations can be cascaded together to provide a concise set of operations. Therefore, the statement for obtaining the even numbers from the vector **v1** can be written in a single line of code.

```
> evens<-v1[v1 %% 2 == 0]
> evens
[1] 2 4
```

R's **which()** function is used to find the location index of vector values. For example, to create a new vector of even numbers it is possible to first find the location the even numbers in the vector, and then use these indices to create the new vector.

```
> v1
[1] 1 2 3 4 5
> ind<-which(v1 %% 2 == 0)
> ind
[1] 2 4
> evens<-v1[ind]
> evens
[1] 2 4
```

This process can be written in a single line of code.

```
> evens<-v1[which(v1 %% 2 == 0)]
> evens
[1] 2 4
```

Indexing can also be used to extract elements from a vector, using the colon operator (:), which generates regular sequences within a specified range. These sequences can be applied to filter the original vector. A minus sign can be used to exclude a range of indices from the calculation.

```
> 2:4
[1] 2 3 4
> v1[2:4]
[1] 2 3 4
> v1[-(2:4)]
[1] 1 5
> v1[-1]
[1] 2 3 4 5
```

The function **seq()** is used to generate a sequence vector in arithmetic progression, and this will be used in the R system dynamics models to setup the simulation time. For example, the vector **times** is a sequence from 0 to 5 (inclusive).

```
> times<-seq(from=0,to=5)
> times
 [1]  0  1  2  3  4  5
```

What is convenient about the **seq()** function is that it can accept an additional parameter (**by**) which can vary the distance between the different elements.

```
> times<-seq(from=0,to=5,by=.5)
> times
 [1] 0.0 0.5 1.0 1.5 2.0 2.5 3.0 3.5 4.0 4.5 5.0
```

Vectors can also be processed using the vectorized **ifelse(b,u,v)** function, which accepts a boolean vector **b** and allocates the element-wise results to be either **u** or **v**. For example, a new character vector can be formed with elements classified as "EVEN" or "ODD" depending on the input vector's value.

```
> ans<-ifelse(v1%%2==0,"EVEN","ODD")
> ans
[1] "ODD"  "EVEN" "ODD"  "EVEN" "ODD"
```

Two additional vectorized functions are useful. These are **all()** and **any()** which process the entire vector and report an overall single condition. It is an efficient form of carrying out a sequence of logical AND (all) or logical OR (any) tests on the vector elements.

```
> v1
[1] 1 2 3 4 5
> any(v1==1)
[1] TRUE
> any(v1<0)
[1] FALSE
> all(v1>=0)
[1] TRUE
```

The elements of a vector can also be allocated names, and in later chapters parameters in a simulation model will be identified this way. Here names are added to the original vector **v1**, and these are then displayed at the console.

```
> v1
[1] 1 2 3 4 5
> names(v1)<-c("a","b","c","d","e")
> v1
a b c d e
1 2 3 4 5
```

A useful feature of naming vector elements is that the name also provides an index to access the value.

```
> v1
a b c d e
1 2 3 4 5
> v1["c"]
c
3
```

Vectors can be increased with new elements. At an implementation level, a new variable is created in memory when a vector is added to, so some computational overhead is involved. This example shows how elements can be added to the end of a vector, using the concatenate (c) function.

```
> v1
[1] 1 2 3 4 5
> v1<-c(v1,c(6,7))
> v1
[1] 1 2 3 4 5 6 7
```

Elements can also be added to the start of a vector.

```
> v1
[1] 1 2 3 4 5
> v1<-c(c(-1,0),v1)
> v1
[1] -1  0  1  2  3  4  5
```

Lists

R's list structure can combine objects of different types. For example, using the **list** () function, a variable is created that can represent information on a student.

```
s<-list(id="1234567",fName="Jane", sName="Smith", age=21)
```

The list variable shows the components of the list (known as tags).

```
> s
$id
[1] "1234567"
$fName
[1] "Jane"
$sName
[1] "Smith"
$age
[1] 21
```

List elements can be accessed through the operator $, for example.

```
> s$fName
[1] "Jane"
> s$age
[1] 21
```

Technically, a list is a vector, and elements it can also be accessed through its index, although double brackets are used instead of single ones to return a vector.

```
> s[[1]]
[1] "1234567"
> s[[2]]
[1] "Jane"
```

Also, elements can be returned using single brackets containing the name of the data type.

```
> s["fName"]
$fName
[1] "Jane"
> s["age"]
$age
[1] 21
```

New elements can be added to a list by simply adding a new element to the variable. The **str()** function can be used to view the structure of an R variable.

```
s$gender<-'F'

> str(s)
List of 5
 $ id     : chr "1234567"
 $ fName  : chr "Jane"
 $ sName  : chr "Smith"
 $ age    : num 21
 $ gender: chr "F"
```

Elements can also be removed from a list, by setting the relevant element to NULL.

```
s$age<-NULL

> str(s)
List of 4
 $ id     : chr "1234567"
 $ fName  : chr "Jane"
 $ sName  : chr "Smith"
 $ gender: chr "F"
```

The list elements can be accessed directly, using the **names()** function.

```
> names(s)
[1] "id"      "fName"   "sName"   "gender"
```

The data contained in a list can be returned as a single vector, using the **unlist()** function. Note that because the vector must contain elements of the same type, the age value is coerced into a character string.

```
> unlist(s)
       id      fName      sName        age     gender
"1234567"     "Jane"    "Smith"       "21"        "F"
```

Finally, interesting things can be done with lists. For instance, they can be recursive, which means a list can contain lists. The earlier example can be extended to do this, by adding an extra student.

```
s1<-list(id="1234567",fName="Jane", sName="Smith", age=21)
s2<-list(id="1234568",fName="Matt", sName="Johnson", age=25)
```

The two lists (representing each individual student) are added to a new list, and this list is then "a list of lists".

```
l<-list(s1,s2)
```

The list output can be summarized as follows, which shows that each element contains a list of 4 elements.

```
> str(l)
List of 2
 $ :List of 4
  ..$ id    : chr "1234567"
  ..$ fName: chr "Jane"
  ..$ sName: chr "Smith"
  ..$ age   : num 21
 $ :List of 4
  ..$ id    : chr "1234568"
  ..$ fName: chr "Matt"
  ..$ sName: chr "Johnson"
  ..$ age   : num 25
```

Matrices

A matrix is a data structure that has a number of rows and columns, where each element has the same mode. Matrix subscripts, similar to vectors, commence at [1,1], and these are used to access row and column elements. A matrix can be initialized from a vector, where the numbers of rows and columns are specified as parameters. R stores matrices by column-major order, and by default matrices are filled in this manner. A matrix can be populated in row-major order by passing the parameter **byrow = TRUE** to the matrix function.

```
> m<-matrix(c(10,20,30,40,50,60),nrow=3,ncol=2)
> m
     [,1] [,2]
[1,]   10   40
[2,]   20   50
[3,]   30   60
```

Matrix elements can be accessed using their row and column numbers as indices.

```
> m[1,1]
[1] 10
> m[3,2]
[1] 60
```

Individual rows can be accessed in a convenient way, by removing the index for a specific column. For this, a vector of row elements is returned.

```
> m
     [,1] [,2]
[1,]   10   40
[2,]   20   50
[3,]   30   60

> m[1,]
[1] 10 40
```

Columns can be extracted by specifying the column index, and the column values are returned in a vector structure.

```
> m[,2]
[1] 40 50 60
```

The function **dim()** can be used to display the matrix dimension, and the functions **nrow()**, **ncol()** provide information on the number of rows and columns.

```
> dim(m)
[1] 3 2
> nrow(m)
[1] 3
> ncol(m)
[1] 2
```

A further useful set of matrix functions is **rowSums()** and **colSums()**, which sum all row and column elements respectively.

```
> rowSums(m)
[1] 50 70 90
> colSums(m)
[1]  60 150
```

In a similar way, the functions **rowMeans()** and **colMeans()** calculate the means of rows and columns.

```
> rowMeans(m)
[1] 25 35 45
> colMeans(m)
[1] 20 50
```

Filtering can also be performed on matrices. For example, if a query is required to find all rows that have *column 1 values greater than 20*, the following code could be used. First a logical vector could be applied to the full column with the specified condition.

```
> test<-m[,1] > 20
> test
[1] FALSE FALSE  TRUE
```

Table 2.1 Useful matrix operations in R

Operator or function	Description
A * B	Element-wise multiplication
A/B	Element-wise division
A %*% B	Matrix multiplication
t (A)	Transpose of A
e<-eigen (A)	List of eigenvalues and eigenvectors for matrix A

This logical vector can then be applied to the row index for the matrix to filter out all FALSE values, and in this case, return the 3rd row, which matches the condition.

```
> m[test,]
[1] 30 60
```

R matrices support linear algebra operations, and this feature will be used in the epidemiology system dynamics model of Chap. 5. Table 2.1 summarizes these operations.

Rows and columns can be added to a matrix, using **rbind()** and **cbind()**, where a vector of appropriately sized values is included as an argument.

```
> rbind(m,c(40,70))
     [,1] [,2]
[1,]   10   40
[2,]   20   50
[3,]   30   60
[4,]   40   70

> cbind(m,c(70,80,90))
     [,1] [,2] [,3]
[1,]   10   40   70
[2,]   20   50   80
[3,]   30   60   90
```

Data Frames

A data frame is similar to a matrix, as it has a two-dimensional rows and columns structure, however it differs from a matrix in that each column can have a different mode (Matloff 2009). This is convenient for data processing, as many real-world data sets consist of tables with different data types, and these can be easily replicated in data frames. For example, the student example presented earlier can be represented in a data frame, by specifying each attribute as a vector, and then combining these into a data frame. The list items were:

```
s1<-list(id="1234567",fName="Jane", sName="Smith", age=21)
s2<-list(id="1234568",fName="Matt", sName="Johnson", age=25)
l<-list(s1,s2)
```

Based on this data, we can identify four different vectors as follows.

```
ids<-c("1234567","1234568")
fNames<-c("Jane","Matt")
sNames<-c("Smith","Johnson")
ages<-c(21,25)
```

These vectors can be combined into a data frame, which represents data similar to the manner in which it is stored in a convention spreadsheet. Attributes are lined up in columns, and each individual observation is stored in a row. The flag **stringsAsFactors** is set to FALSE, which means R will not convert strings to factors, which are used to represent categorical variables in R.

```
s<-data.frame(ID=ids,FirstName=fNames,Surname=sNames,
              Age=ages,stringsAsFactors=FALSE)

> s
       ID FirstName Surname Age
1 1234567      Jane   Smith  21
2 1234568      Matt Johnson  25
```

Technically, a data frame is a list, and so the list notation can be used to access information. For example, columns can be accessed using the double bracket notation [[]], and individual elements can also be extracted from columns by applying a further index to locate the value.

```
> s[[1]]
[1] "1234567" "1234568"
> s[[1]][1]
[1] "1234567"
```

A data frame can also be accessed using matrix operators, where the structure is accessed via its rows and columns.

```
> s[1,]
       ID FirstName Surname Age
1 1234567      Jane   Smith  21
> s[,1]
[1] "1234567" "1234568"
> s[1,1]
[1] "1234567"
```

Finally, data frames elements can be accessed using the column names as follows.

```
> s$Surname
[1] "Smith"   "Johnson"
```

Filtering can be performed by applying conditional statements to the data frame, for example, finding all students whose age is greater than 21.

```
> s[s$Age > 21,]
       ID FirstName Surname Age
2 1234568      Matt Johnson  25
```

This query can also be applied using the **subset()** function, which takes a data frame and applies a filtering condition.

```
> sb<-subset(s,s$Age>21)
> sb
       ID FirstName Surname Age
2 1234568      Matt Johnson  25
```

Additional columns can be conveniently added to a data frame. For example, if all students under the age of 21 were eligible for a discount, the following command would add this information as a new column in the data set.

```
> s$Discount<-ifelse(s$Age<=21,"YES","NO")
> s
       ID FirstName Surname Age Discount
1 1234567      Jane   Smith  21      YES
2 1234568      Matt Johnson  25       NO
```

For data analysis, opportunities often arise by merging different data sets, and the **merge()** function facilitates this. In the student example, a second data frame could store examination results for each student.

```
ids<-c("1234567","1234568")
subjects<-c("CT111","CT111")
grade<-c(80,80)

r<-data.frame(ID=ids,Subject=subjects,Grade=grade,
              stringsAsFactors=FALSE)
> r
       ID Subject Grade
1 1234567   CT111    80
2 1234568   CT111    80
```

As this data frame shares a common attribute with the student information (i.e. the ID value), the two data frames can be merged based on this column (passed as an argument to the merge function).

```
> new<-merge(s,r,by="ID")
> new
       ID FirstName Surname Age Subject Grade
1 1234567      Jane   Smith  21   CT111    80
2 1234568      Matt Johnson  25   CT111    80
```

The merged data frame could then be used to support statistical analysis of a large data set, for example, to test whether there is a link between factors such as age, and examination performance.

Functions

A function is a group of instructions that takes input, uses the input to compute values, and returns a result (Matloff 2009). Users of R should adopt the habit of creating simple functions which will make their work more effective and also more trustworthy (Chambers 2008). Functions are declared using the **function** reserved word. They contain a list of parameters (some of which may have default values), and execute a set of instructions between an opening brace ({) and a closing brace (}).

```
convC2F<-function(celsius)
{
  fahr<-celsius*9/5 + 32.0
  return(fahr)
}

> convC2F(100)
[1] 212
```

This initial function converts temperature in Celsius to its corresponding value in Fahrenheit. The formal parameter **celsius,** and the variable **fahr** are both local to the function, which means they are no longer available after the function completes its task. This is important, as it enables information hiding within the function, and ensures that all direct communication between functions is done through its arguments and return value. Variables declared outside of functions are global, and are visible within the functions. The second function shows how a loop structure can be used within the function in order to calculate the result—in this case to count the frequency of even numbers in a vector. Notice that the return statement is omitted, as the last evaluated expression is by default returned by functions in R, and avoiding a return statement can improve code performance.

```
evenCount<-function(v)
{
  ans<-0
  for(x in v)
  {
    if(x%%2==0)
      ans<-ans+1
  }
  ans # more efficient method for returning values
}
```

The function is tested by passing in an arbitrary vector, and observing the result.

```
> evenCount(c(2,2,1,2))
[1] 3
```

Apply Functions

Another use of user-defined functions in R is as a parameter to the *apply* family of functions, which are one of the most famous and used features of R (Matloff 2009). The general form of the **sapply(x,f,fargs)** function is as follows:

- **x** is the target vector or list
- **f** is the function to be called
- **fargs** are the optional set of arguments that can be applied to the function **f**.

The **sapply()** function takes as input a target vector and a function. The function specifies the logic that is executed on each vector element, and **sapply()** then returns a vector with the processed data. For example, *if there was a requirement to calculate the difference between each value in a vector and the overall vector mean, the following code could be used.*

First, the sample **data** is generated, with 10 random values between 1 and 10, using the function **sample()**, where replacement is enabled. The mean is calculated using the **mean()** function.

```
> data<-sample(1:10,replace=T)
> data
 [1]  9  2  8 10  9  1  8  2  1  6
> mean(data)
[1] 5.6
```

This **sapply()** call to perform this task, shown below, takes three parameters:

- The vector to be iterated over, which is the vector **data**.
- The function to process each element. This function is declared within the **sapply** call itself, and takes two parameters, **e** and **m**. The parameter **e** is the current vector element being processed, and the parameter **m** is the vector mean. The function then evaluates the difference between the two values, and this is processed by **sapply** and a vector returned after all the elements have been processed.
- The third parameter maps onto the second argument (m) to be passed to the function, which is the mean of the vector.

```
> d<-sapply(data,function(e,m){e-m}, mean(data))
```

The resulting vector displays the difference between each element and the overall vector mean.

```
> d
 [1]  3.4 -3.6  2.4  4.4  3.4 -4.6  2.4 -3.6 -4.6  0.4
```

The *apply* functions can also be used to process lists, as well as vectors. For example, consider the following list of students.

```
s1<-list(id="1234567",fName="Jane", sName="Smith", age=21)
s2<-list(id="1234567",fName="Matt", sName="Johnson", age=25)
l<-list(s1,s2)
```

The task here is to implement a simple query: find the list elements (in the list **l**) whose age is greater than 21. This can be done in two steps. First, **sapply()** is used to process the query and return a boolean vector indicating the list indices that match the conditional expression, and the result is stored in the vector **b**.

```
> b<-sapply(l,function(x)x$age>21)
> b
[1] FALSE  TRUE
```

Next, the vector b can be used to filter the original list, and the answer is stored in the **ans**, which now contains all those elements that match the condition.

```
> ans<-l[b]
> str(ans)
List of 1
 $ :List of 4
  ..$ id   : chr "1234568"
  ..$ fName: chr "Matt"
  ..$ sName: chr "Johnson"
  ..$ age  : num 25
```

The **apply()** function can be used to process rows and columns for a matrix, and the general form of this function (Matloff 2009) is **apply(m, dimcode, f, fargs)**, where:

- **m** is the target matrix
- **dimcode** identifies whether it's a row or column target. The value 1 is used to process rows, whereas 2 applies to columns
- **f** is the function to be called
- **fargs** are the optional set of arguments that can be applied to the function **f**.

For example, **apply()** can be used to find the mean value in each row.

```
> m
     [,1] [,2]
[1,]   10   40
[2,]   20   50
[3,]   30   60
> apply(m,1,mean)
[1] 25 35 45
```

In a similar way, **apply()** can be used to find the mean value in each column.

```
> apply(m,2,mean)
[1] 20 50
```

deSolve Package

R's **deSolve** package solves initial value problems written as ordinary differential equations (ODE), differential algebraic equations (DAE), and partial differential equations (PDE) Soetaert et al. (2010). For system dynamics models, the ODE solver in **deSolve** is used. The key requirement is that system dynamics modelers implement the model equations in a function, and this function is called by **deSolve.** For this example the customer growth model from Chap. 1 is revisited, as shown in Fig. 2.1.

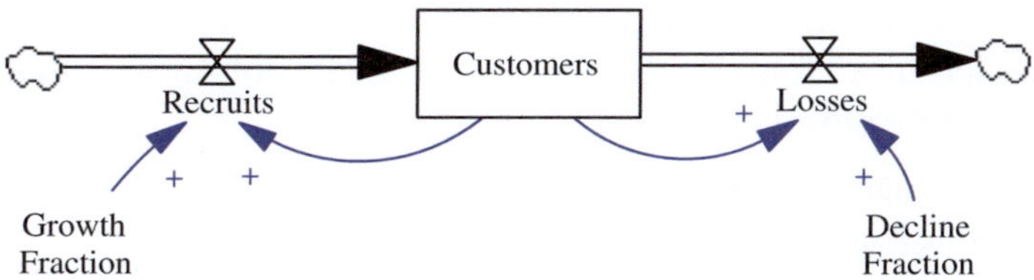

Fig. 2.1 A stock and flow model of customers (from Chap. 1)

The R implementation of this model is now described. To use the **deSolve** library, the package needs to be installed, and then it should be referenced in the model source file by calling the **library()** function.

```
library(deSolve)
```

In the R implementation, the first task is to define the simulation time constants, and then create the simulation time vector using the **seq()** function.

```
START<-2015; FINISH<-2030; STEP<-0.25
simtime <- seq(START, FINISH, by=STEP)
```

The vector **simtime** can be inspected, and it is useful to see how the **seq()** function creates the list of times from start to finish, with the appropriate steps in between. The **head()** and **tail()** function are used to display the first and final six elements of the vector.

```
> head(simtime)
[1] 2015.00 2015.25 2015.50 2015.75 2016.00 2016.25
> tail(simtime)
[1] 2028.75 2029.00 2029.25 2029.50 2029.75 2030.00
```

Next, two model vectors must be defined, as these are required as inputs to the system dynamics model function. The first vector is named **stocks** and contains the model stocks, along with their initial values. For this example, there is only a single stock, and its initial value is set to 10000. To improve model readability, a computer programming convention known as Hungarian notation is used to prefix a variable name with it system dynamics type, i.e. *s* for stock, *f* for flow and *a* for auxiliary).

```
stocks  <- c(sCustomers=10000)
```

The second vector is called **auxs** and this contains the exogenous parameters for the customer model.

```
auxs    <- c(aGrowthFraction=0.08, aDeclineFraction=0.03)
```

When simulating with **deSolve**, the modeler must write a function to implement the model equations. The user-defined function, arbitrarily named **model()**, and called from the **deSolve** library, takes three parameters:

- The current simulation time (*time*),
- A vector of all current stock values (*stocks*).
- A vector of model parameters (*auxs*).

These vectors can be transformed to lists using **as.list()**, and embedded in the **with()** function, as this allows the variable names to be conveniently accessed.

```
model <- function(time, stocks, auxs){
  with(as.list(c(stocks, auxs)),{

    fRecruits<-sCustomers*aGrowthFraction

    fLosses<-sCustomers*aDeclineFraction

    dC_dt <- fRecruits - fLosses

    return (list(c(dC_dt),
                 Recruits=fRecruits, Losses=fLosses,
                 GF=aGrowthFraction,DF=aDeclineFraction))
  })
}
```

With these input values, all that remains is to specify the stock and flow equations in their correct solving sequence.

- The flow *fRecruits* is a product of the stock *sCustomers* and the growth fraction *aGrowthFraction.*
- The flow *fLosses* is a product of the stock *sCustomers* and the decline fraction *aDeclineFraction.*
- The net flow (derivative) for the stock is calculated as the difference in inflow and outflow, and stored in the variable *dC_dt.*

A list structure is then returned to the **deSolve** package. The first parameter is a vector of all the net flows, and this *must match the order in which the stocks are initialized* in the vector **stocks**. Following this, any other model variable can be added to the return list to ensure that appears as part of the final result set. In this case, the flows and auxiliaries are added, and user-friendly names provided.

Finally, the model is solved by calling the **ode()** function, which is part of the **deSolve** library. This function takes five arguments.

- The vector of stocks (*y=stocks*).
- The simulation time vector (*times=simtime).*
- The function name that contains the model equations (*func = model).*
- The auxiliary parameters (*parms=auxs).*

- The integration method *(method="euler").* Other methods are available, including Runge-Kutta 4th order integration (*method="rk4").*

```
o<-data.frame(ode(y=stocks, times=simtime, func = model,
                  parms=auxs, method="euler"))
```

The full set of simulation results from **ode** are then converted into a data frame, and using R's **head()** function, the first six rows of results are displayed.

```
> head(o)
     time sCustomers Recruits   Losses   GF   DF
1 2015.00   10000.00 800.0000 300.0000 0.08 0.03
2 2015.25   10125.00 810.0000 303.7500 0.08 0.03
3 2015.50   10251.56 820.1250 307.5469 0.08 0.03
4 2015.75   10379.71 830.3766 311.3912 0.08 0.03
5 2016.00   10509.45 840.7563 315.2836 0.08 0.03
6 2016.25   10640.82 851.2657 319.2246 0.08 0.03
```

This data frame can be used as a basis to plot data and also to analyze results. For example, the **summary()** function can be applied to the stock and flows in the data frame, yielding useful summary statistics (columns 1, 5 and 6 are omitted).

```
> summary(o[,-c(1,5,6)])
   sCustomers          Recruits            Losses
 Min.   :10000    Min.   : 800.0    Min.   :300.0
 1st Qu.:12048    1st Qu.: 963.9    1st Qu.:361.4
 Median :14516    Median :1161.3    Median :435.5
 Mean   :14866    Mean   :1189.3    Mean   :446.0
 3rd Qu.:17489    3rd Qu.:1399.2    3rd Qu.:524.7
 Max.   :21072    Max.   :1685.7    Max.   :632.2
```

Visualization

R provides visualization libraries, and throughout this text, the R package **ggplot2** is used. The terminology used in ggplot2 (Chang 2013) includes:

- The *data* to be visualized, which consists of variables stored in a data frame.
- The geometric objects, or *geoms*, that are drawn to represent the data, such as points and lines.
- Aesthetic attributes that are the visual properties of geoms, such as line color, point shape etc.

As the simulation results are already in a data frame, they are ready to be visualized by calls to **ggplot2**. The package **ggplot2** must be included, and following that a call to the function **ggplot()** is executed. A *layered approach* to building a plot is used, with where additional attributes are added in sequence, following the first call to **ggplot()**. With this example, a line is added specifying the data frame, and the function **aes()** specifies the *x* and *y* variables.

```
library(ggplot2)

ggplot()+
  geom_line(data=o,aes(time,o$sCustomers),colour="blue")+
  geom_point(data=o,aes(time,o$sCustomers),colour="blue")+
  scale_y_continuous(labels = comma)+
  ylab("Customers")+
  xlab("Year")
```

Figure 2.2 shows the variable of interest (s*Customers*) changing over time. Additional variables can be added to the plot by adding further calls to **geom_line()**,

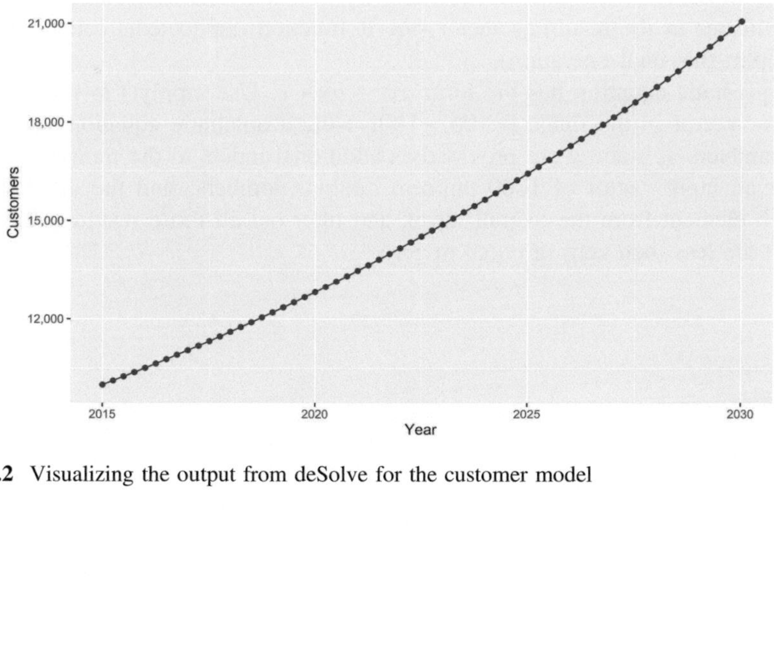

Fig. 2.2 Visualizing the output from deSolve for the customer model

and this data is also be presented in point format by using the function **geom_point()**. High resolution plots can also be created to support publication-quality presentations. Following the **ggplot()** call, the function **ggsave()** will save the image to a file on the disk, and this supports a range of formats.

```
ggsave("customers.png")
```

In summary, **ggplot2** is a powerful visualization framework. A more comprehensive listing of its features is outside the scope of this text, however Chang (2013) provides an excellent source of examples that can be built upon to maximize the visualization impact of simulation output.

Summary

In conclusion, R is a powerful data analytics platform that supports system dynamics modeling through the **deSolve** package. Further benefits from using R is the facility to vectorize simulation models, analyze data, and apply further statistical analysis to simulation output. For example, Chap. 5 will show how disaggregate system dynamics models can be created using R. Chapter 6 demonstrates how R's unit testing framework can be used to test models, and Chap. 7 provides examples of model calibration, sensitivity analysis, and statistical screening, that can all be used to enhance the model building process.

Exercises

1. Create a vector of 100 random numbers, in the range 1–10. From this vector, filter those variables that are divisible by 2. Finally, ensure that there are no duplicates in the resulting vector (the R function duplicated() can be used to support this final operation).
2. A quadratic equation has the form $ax^2 + bx + c$. Use sapply() to transform an input vector in the range [−100, +100] using a quadratic equation, where the parameters a, b and c are provided as additional inputs to the transformation.
3. For an input vector of 1000 uniform random numbers, find the difference of each element from the overall mean, and filter out all those resulting elements that are less than zero or equal to zero.

References

Chambers J (2008) Software for data analysis: programming with R. Springer Science & Business Media, Chicago
Chang W (2013) R graphics cookbook. O'Reilly Media Inc., Sebastapol, CA
Matloff N (2009) The art of R programming. No Starch Press, San Francisco, CA
Soetaert KER, Petzoldt T, Setzer RW (2010) Solving differential equations in R: package deSolve. J Stat Soft 33

Chapter 3
Modeling Limits to Growth

There will always be limits to growth. They can be self-imposed. If they aren't, they will be system-imposed.
Donella H. Meadows, Thinking in Systems: A Primer (2008, p. 103).

Abstract This chapter introduces system dynamics models of *limits to growth*. First, a *one-stock* model is presented, where the growth rate varies, and is influenced by the system's carrying capacity. Second, a model of economic growth is described, which captures the law of diminishing returns, a feature of many economic systems. Third, a *two-stock* model of limits to growth is specified, where a growing stock consumes its carrying capacity, and this dynamic leads to growth followed by rapid decline. Before introducing the limits to growth models, an explanation of an important formulation method in system dynamics is presented. This allows modelers to construct robust equations to model the effect of one variable on another. This is useful for many system dynamics models, particularly where one system stock influences another system stock.

Keywords Formulating effects · Limits to growth · Economic growth · Overshoot and collapse

Modeling Causal Relationships Using Effects

An important building block for system dynamics is to model how variables influence one another over time. While some of these may be simple linear relationships, the reality is that real-world effects between variables can also be non-linear, and may also involve multiple variables. System dynamics offers a convenient structure for modeling the effect of one variable on another (Sterman 2000), and this approach is described in (3.1) and (3.2).

$$Y = Y^* \times Effect(X_1\ on\ Y) \times \cdots \times Effect(X_n\ on\ Y) \tag{3.1}$$

J. Duggan, *System Dynamics Modeling with R*, Lecture Notes in Social Networks,
DOI 10.1007/978-3-319-34043-2_3

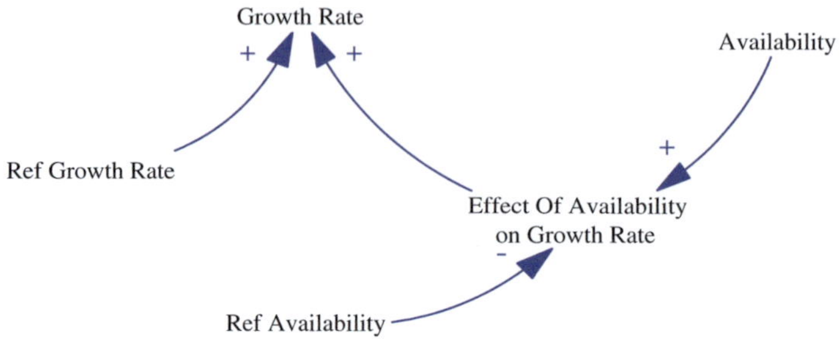

Fig. 3.1 Formulating the effect of availability on growth rate

$$Effect(X_i \; on \; Y) = f(X_i/X_i^*) \tag{3.2}$$

These two equations are based on the following assumptions.

- Variable Y is the dependent variable of a causal relationship, and this is a function of n independent variables (X_1, X_2, ..., X_n)
- Variable Y has a reference value Y^*, which is the normal value the variable Y takes on. This reference value is multiplied by a sequence of *effect functions* that are calculated based on the normalized ratio of input term (X_i/X_i^*), where X_i^* is the reference input value, and X_i is the actual input value.
- The effect function has the normalized ratio (X/X^*) on its x-axis, and always contains the point (1, 1). This point (1, 1) is important for the following reason: if X equals its reference value X^*, then the effect function will be 1, and therefore Y will then equal its reference value Y^* (from Eq. 3.1).

Consider the following model, which will form part of this chapter's first limits to growth model (Fig. 3.1).

In this example, based on the generic formulation (3.1) with just a single X value, the growth rate of a system (3.3) is the product of the reference growth rate (3.5) times the *effect of resource availability* on the growth rate (3.4). Therefore, if the availability equals the reference availability, the effect will be 1, and the growth rate will equal the reference growth rate.

$$Growth\ Rate = Ref\ Growth\ Rate \times Effect\ of\ Availability\ on\ Growth\ Rate \tag{3.3}$$

$$Effect\ of\ Availability\ on\ Growth\ Rate = f\left(\frac{Availability}{Ref\ Availability}\right) \tag{3.4}$$

$$Ref\ Growth\ Rate = 0.10 \tag{3.5}$$

$$Ref\ Availability = 1.0 \tag{3.6}$$

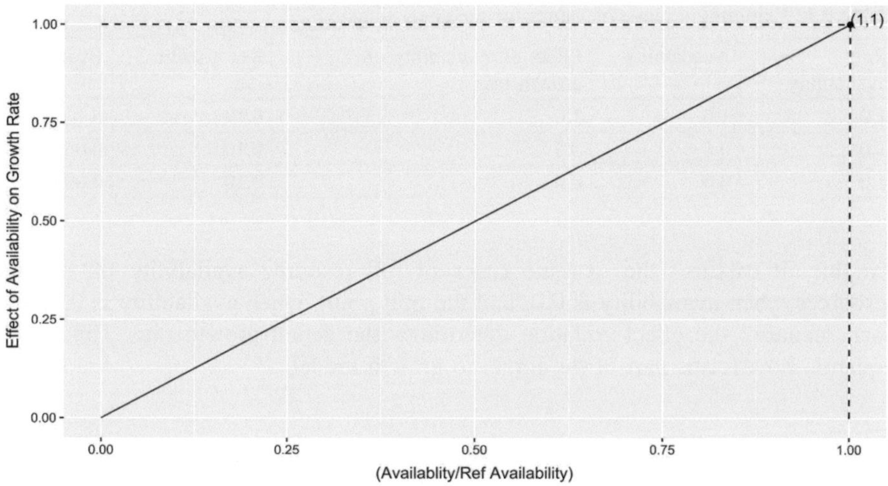

Fig. 3.2 The relationship between availability and the effect on growth rate

The effect Eq. (3.4) is now explored in further detail. The concept is straightforward, and the two extreme cases can be considered in order to explore the relationship between availability and growth rate.

- If availability is 1 (its maximum possible value), then the effect is 1, and the growth rate will take on its maximum value.
- If availability is zero, for example, there are no resources to support further growth, then the effect is zero, and the growth rate is therefore zero.

To keep the model simple, the assumption is that the relationship between availability and growth rate is linear, and this is illustrated in Fig. 3.2. On the x-axis is the dimensionless ratio of availability to reference availability, and the y-axis contains the related effect value.

The algebraic equation of a line with slope m and intercept c is:

$$y = mx + c$$

In this case, the intercept c = 0, and the slope, $m = (y_2 - y_1)/(x_2 - x_1)$, is 1. Elaborating on 3.4, our effect equation is now represented by Eq. 3.7.

$$\textit{Effect of Availability on Growth Rate} = \frac{\textit{Availability}}{\textit{Ref Availability}} \tag{3.7}$$

To highlight possible growth rate values, consider Table 3.1, which shows how the *growth rate* changes depending on the value of *availability*—through the effect

Table 3.1 Exploring a range of values for the effect function

Ref availability	Availability	Effect of availability on growth rate	Ref growth rate	Growth rate
1.0	1.0	1.0	0.10	0.10
1.0	0.5	0.5	0.10	0.05
1.0	0.0	0.0	0.10	0.00

variable. It captures the extreme cases of full resource availability of 1.0, no resources when availability is 0.0, and the mid-point, when availability is 0.5. For each scenario, the effect equation determines the actual growth rate. This effect equation now forms part of the limits to growth model.

S-Shaped Growth

Meadows (2008) writes that there will always be limits to system growth, and that these can be self-imposed, or, failing that, imposed by the system. For instance, market saturation for a product is an example of a limit to growth, as the potential adopters are converted to adopters until (in theory) there are no potential adopters remaining. The spread of a virus is similar, as people change state from being susceptible to infected, and the limit for the virus spread is the total number of susceptible people in the population.

Earlier in chapter one, a one-stock feedback model of capital growth was presented. A similar one-stock structure is now described, but a new feature is added. This model introduces a limiting factor, which acts as a balancing loop that counteracts growth. From Fig. 3.3, the model contains the following elements.

- The *stock* (3.8) has a single inflow, with an initial value of 100 units.
- This inflow (3.9) is the product of the *growth rate* and the stock, where the growth rate is defined in (3.3).
- The growth rate is a product of the reference growth rate and the effect of availability on the growth rate. These variables have already previously defined in (3.5), (3.6) and (3.7).
- Availability is a ratio that measures how much capacity remains in the system, and it is specified in (3.10). The capacity is an arbitrary constant value (3.11). When the stock equals the capacity, availability is zero, and this ensures that there is no further growth in the system.

$$Stock = INTEGRAL(Net\ Flow, 100) \tag{3.8}$$

$$Net\ Flow = Stock \times Growth\ Rate \tag{3.9}$$

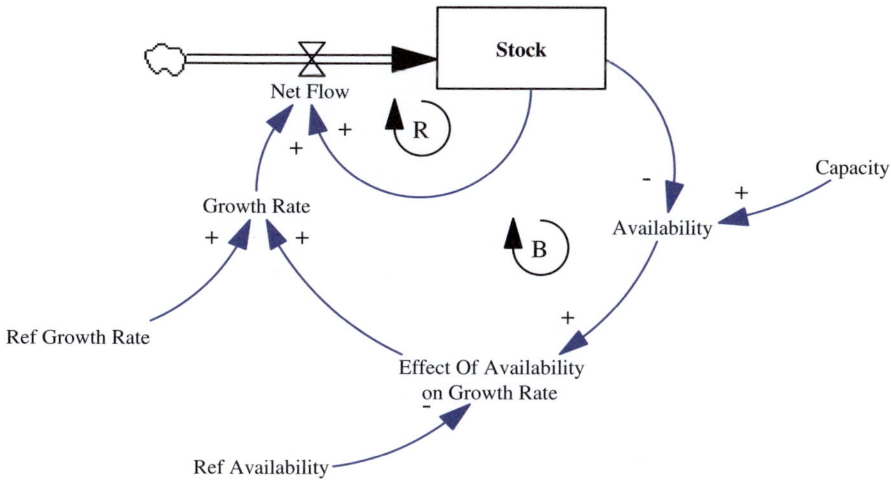

Fig. 3.3 A one-stock model of limits to growth

$$Availability = 1 - \frac{Stock}{Capacity} \tag{3.10}$$

$$Capacity = 10000 \tag{3.11}$$

With this model specification complete, the implementation in R is described. The first task is to include the three libraries, **deSolve** for the numerical integration, and **ggplot2** and **gridExtra** for visualizing the output.

```
library(deSolve)
library(ggplot2)
library(gridExtra)
```

Following this, a number of variables are declared that will be used for the simulation run. These include:

- The start time, finish time and simulation step.
- A time vector **simtime**, which contains the sequence of times where the solver must solve for the variables.
- A vector **stocks** for the system stocks, and this must include the initial values for all stocks (3.8).
- A vector **auxs** that contains a list of the auxiliary constants for the model. In this case, included are the capacity (3.11), the reference availability (3.6) and the reference growth rate (3.5).

```
START<-0; FINISH<-100; STEP<-0.25

simtime    <- seq(START, FINISH, by=STEP)
stocks     <- c(sStock=100)
auxs       <- c(aCapacity=10000,
                aRef.Availability=1,
                aRef.GrowthRate=0.10)
```

The model equations are embedded into the R function **model**. This function will be called by the **deSolve** library for each timestep, where each invocation passes in the current time, a vector of stocks with their current simulation values, and a vector of auxiliaries. From these values, the equations are evaluated in the required sequence, starting with availability (3.10), the effect function (3.7), the growth rate (3.3), and the net flow (3.9). The integral (3.8) is represented by the variable **dS_dt**, and this is returned in the first element of the list (a vector, as there may be more than one stock in a model). The remaining list elements contain other model variables that will be added to the simulation output.

```
model <- function(time, stocks, auxs){
  with(as.list(c(stocks, auxs)),{
    aAvailability <- 1 - sStock / aCapacity
    aEffect <- aAvailability / aRef.Availability
    aGrowth.Rate <- aRef.GrowthRate * aEffect
    fNet.Flow <- sStock * aGrowth.Rate
    dS_dt  <- fNet.Flow
    return (list(c(dS_dt),NetFlow=fNet.Flow,
                 GrowthRate=aGrowth.Rate, Effect=aEffect,
                 Availability=aAvailability))
  })
}
```

The output from **ode** is converted to a data frame, and plotted (Fig. 3.4) using the **qplot()** function.

The plots in Fig. 3.4 can be explored, in a clockwise direction, to gain insight into the workings of this model.

- The first plot, the *stock*, exhibits s-shaped growth behavior, which is the classic mode for a limits to growth model. This is characterized by exponential growth in the early phase. However, shortly after time 46, there is a point of inflection, where curve behavior changes to logarithmic growth, and its value then approaches the limit by time 100.
- The second plot, based on Eq. (3.10) displays the *availability*, and this shows a mirror image of the system stock. Availability is highest when the stock is at its lowest value, and its value falls as the stock rises, and finishes at zero.

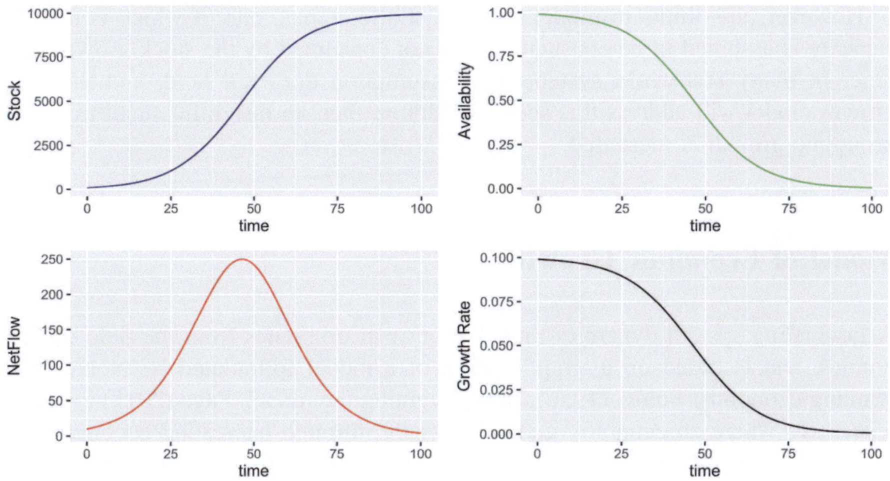

Fig. 3.4 Simulation output of limits to growth model

- The third plot, the *growth rate*, which is based on Eq. (3.3), and is driven by the effect function specified in Eq. (3.7), commences close to its maximum value of 0.10, and then decreases as the availability declines. When the system reaches it fixed capacity, the growth rate drops to zero, and therefore no further growth is possible in the system.
- The fourth plot captures the *net flow*, and this follows a classic bell-shaped growth, where the rate of change increases exponentially, before peaking, and then declining until it reaches zero. This net flow then drives the stock's value.

This system dynamics model is of historical significance (Richardson 1991), and was proposed by the Belgian mathematician Verhulst (1845, 1847). Verhulst noted that population increase is limited by the size and fertility of the country, with the result that the population gets ever-closer to a steady state value. He proposed the following (and somewhat arbitrary) differential equation of the population $P(t)$ at time t:

$$\frac{dP}{dt} = rP\left(1 - \frac{P}{K}\right) \tag{3.12}$$

Equation (3.12) is similar to the net flow equation (3.9). With a growth rate r and limit K, Verhulst went on to compare his results to empirical data in the populations of France, Belgium, the county of Essex in England, and Russia, and the models reported a good fit to the data (Bacaër 2011).

However, this limits to model also has shortcomings, one of which is that the capacity is assumed to be constant, and is not consumed by the stock growth over time. In many real-world systems this assumption does not hold, and the third growth model will address this scenario. Before that, an insightful small model of economic growth is presented.

Model of Economic Growth

A fascinating system dynamics model of growth originates from the field of economics, and is based on a simplification of a model formulated by Nobel Prize winning economist Solow (1956). In highlighting the model, Page (2015) imagines a scenario where an economy generates wealth through harvesting coconuts with machines. A fixed percentage of these resources are reinvested to produce more machines, and therefore increase the economic output. Figure 3.5 shows the model structure, which has one stock with an inflow and outflow.

There are with two feedback loops that interact to drive system behavior, one reinforcing and the other balancing. The reinforcing loop, which is similar to the capital model presented in chapter one, is summarized in Table 3.2.

The loop dynamics determine that the direction of change for the stock is amplified after one loop iteration:

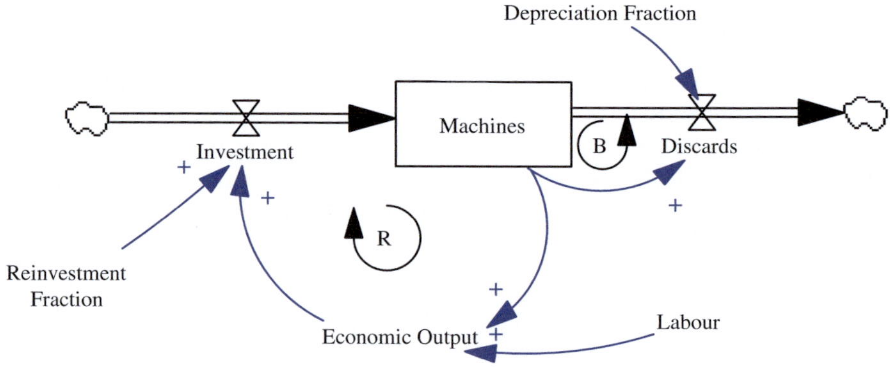

Fig. 3.5 A simple economic model of growth

Table 3.2 Positive feedback loop for the economic model

↑	Machines	→	Economic output	↑
↑	Economic output	→	Investment	↑
↑	Investment	→	Machines	↑

Table 3.3 Negative feedback loop for the economic model

↑	Machines	→	Discards	↑
↑	Discards	→	Machines	↓

- As the stock (machines) increase, so to does economic output, as there is more capacity to harvest the resource.
- An increase in economic output leads to an increased in investment.
- As the inflow investment increases, the stock of machines grows.

The second feedback loop is described in Table 3.3. This is a balancing loop, as can be seen when tracing the change in the stock through the loop elements, as the direction of the stock is changed after one loop iteration.

- As machines increase, the number of discards also increase.
- An increase in discards reduces the machine stock.

Following the feedback loop polarity calculation, the model equations are formulated. The stock is specified in (3.13), with an initial value of 100 machines. The inflow (3.14) is the product of the reinvestment fraction (3.16) and the economic output. The outflow (3.15) is a depreciation measure that reduces the stock of machines by a constant fraction (3.17) for each time unit.

$$Machines\ (M) = INTEGRAL\ (Investment - Disards, 100) \tag{3.13}$$

$$Investment = Economic\ Output \times Reinvestment\ Fraction \tag{3.14}$$

$$Discards = Machines \times Depreciation\ Fraction \tag{3.15}$$

$$Reinvestement\ Fraction\ (R) = 0.20 \tag{3.16}$$

$$Depreciation\ Fraction\ (D) = 0.10 \tag{3.17}$$

An important model equation is the economic output (3.18), and this is based on a fundamental of economics. This equation is a convenient model of *diminishing returns,* because the rate of increase in productivity decreases as additional machines are added. This is captured mathematically by using the square root function, which is a widely used concave function (i.e. where the slope is decreasing).

$$Economic\ Output\ (O) = Labour \times \sqrt{Machines} \tag{3.18}$$

$$Labour\ (L) = 100 \tag{3.19}$$

These seven equations are now implemented as a system dynamics model in R. First, the simulation time, stock (with initial value) and constant auxiliaries are defined in the usual vector format, and the model equations are encapsulated in the **model** function, which is called by the **ode** function.

```
START   <-0; FINISH<-100; STEP<-0.25
simtime <- seq(START, FINISH, by=STEP)
stocks  <- c(sMachines=100)
auxs    <- c(aDepFraction=0.1, aLabour=100,
             aReinvestFraction=0.20)

model <- function(time, stocks, auxs){
  with(as.list(c(stocks, auxs)),{
    aEconomicOutput <- aLabour * sqrt(sMachines)
    fInvestment     <- aEconomicOutput * aReinvestFraction
    fDiscards       <- sMachines * aDepFraction
    dM_dt           <- fInvestment - fDiscards

    return (list(c(dM_dt),
                 Investment=fInvestment, Discards=fDiscards,
                 EconomicOutput=aEconomicOutput))
  })
}
o<-data.frame(ode(y=stocks, times=simtime, func = model,
                  parms=auxs, method="euler"))
```

The output is captured in displayed in Fig. 3.6. What is of interest is that over time the stock of machines converges to a constant value, even though more machines are being added. Therefore the marginal benefit, in terms of economic output, of adding new machines decreases until it reaches zero. This is due to the impact of discards, which is the balancing feedback loop in the model.

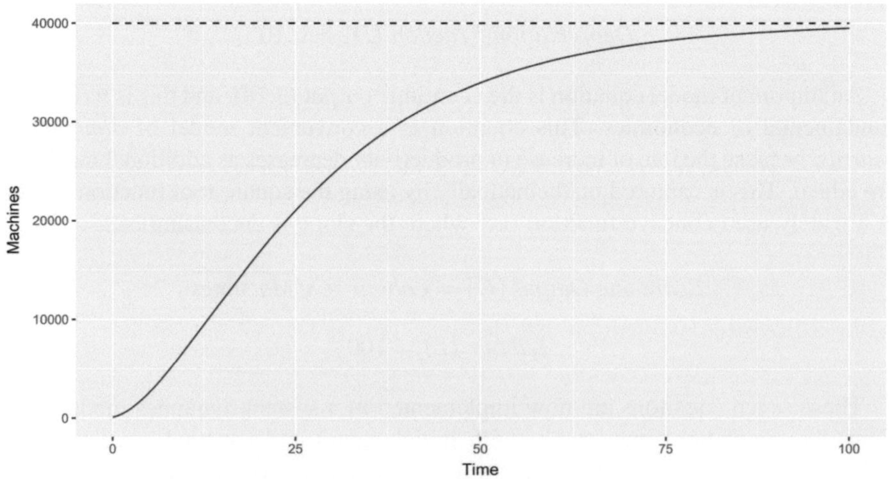

Fig. 3.6 Model output showing limit to growth for machines

Interestingly, if the discard rate was set to zero, the balancing loop would be deactivated, and the number of machines would grow exponentially. However, with the balancing loop active, as the number of machines rise, so too does the discard rate, and over time the model reaches an equilibrium point where the discards equal the investment. When this happens, the machine level is constant (i.e. a dynamic equilibrium), and so economic output also remains constant.

System dynamics also provides the capability to perform equilibrium analysis for this model. A basic principle of system dynamics is that, under equilibrium conditions for any stock, the sum of all inflows will equal the sum of all outflows. This relationship between inflow (3.14) and outflow (3.15) is represented in Eq. (3.20), and rearranged to show the value for M^* in equilibrium (3.21). Interestingly, with L and D, constant, this Eq. (3.21) demonstrates that the number of machines increases with the square of the reinvestment rate. This shows that economic output, which is a function of the square root of the machines (3.18), will only increase linearly as the investment fraction increases.

$$R \times L \times \sqrt{M^*} = M^* \times D \tag{3.20}$$

$$M^* = \left(\frac{RL}{D}\right)^2 \tag{3.21}$$

The actual equilibrium value M can be calculated with values of $R = 0.2$, $L = 100$ and $D = 0.1$, and this is 40,000, which is the same result as the steady state value computed in the simulation model, and shown in Fig. 3.6. While this basic growth model finally generates a fixed level of output (i.e. growth halts), the model is useful as it models growth that results from exploiting a technology to its limits (Page 2015). As such, this initial model does not cater for future innovations in machine technology (e.g. increased productivity and longer life-span), and Solow's more detailed model accommodates this, and so can be used to model further increases in growth (Solow 1956).

Modeling Constraints—A Non-renewable Stock

Continuing with an economic theme, a two-stock system is presented where the growth of one stock depends on the level of a second stock, which is non-renewable. This model explores how the growth of oil wells is ultimately constrained by the availability of oil. The model shows an initial growth phase, as the underlying resource is abundant, followed by a sharp decline as the resource is consumed and depleted. This two-stock example, based on Meadows (2008, p. 60), and shown in Fig. 3.7, focuses on a system that generates revenue through the extraction of a non-renewable resource.

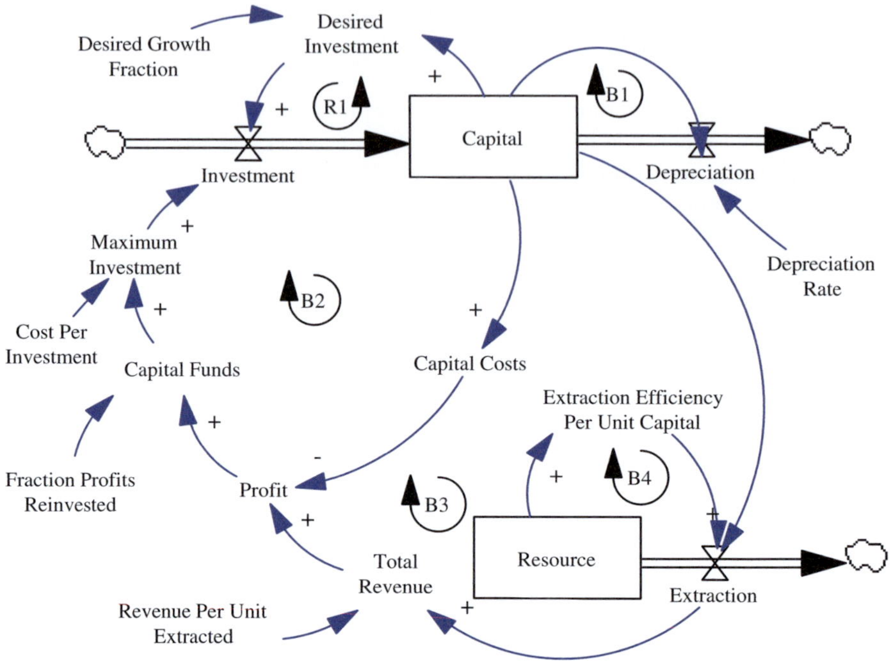

Fig. 3.7 Limits to growth for capital, constrained by a non-renewable resource

This stock and flow model shows that systems with limits to growth have a reinforcing loop driving the growth, and a counteracting balancing loop that constrains growth. The model captures the growth and decline dynamics of a company discovering a new oil field, where the stock of oil could potentially last for up to 200 years. The key features of the model are:

- The capital stock (e.g. oil wells) provides the capability to extract the resource. Investment is needed in capital stock, because equipment degrades over time, and must be replaced. The investment rate is initially determined by the growth goal, but this investment rate is impacted as the resource depletes, which results in limits to further growth.
- The resource stock is non-renewable, which features a single outflow, as it can only be consumed. Resource extraction is based on the amount of available capital. However, extraction rates are impacted by the amount of the available resource. As the resource level drops, the amount of resource extracted per unit capital declines. In the case of oil, this is an important dynamic. As oil resource becomes more dilute, there is less natural pressure to force it to the surface, and therefore more costly and technically sophisticated measures are required for successful extraction (Meadows 2008).

The positive feedback loop (R_1) is summarized in Table 3.4, and this shows an exponential growth process, whereby higher capital leads to further investment, and

Table 3.4 Positive feedback loop (R_1)—capital growth

↑	Capital	→	Desired investment	↑
↑	Desired investment	→	Investment	↑
↑	Investment	→	Capital	↑

Table 3.5 Negative feedback loop (B_1)—capital depreciation

↑	Capital	→	Depreciation	↑
↑	Depreciation	→	Capital	↓

Table 3.6 Negative feedback loop (B_2)—increasing capital, increasing costs

↑	Capital	→	Capital costs	↑
↑	Capital costs	→	Profit	↓
↓	Profit	→	Capital funds	↓
↓	Capital funds	→	Maximum investment	↓
↓	Maximum investment	→	Investment	↓
↓	Investment	→	Capital	↓

in turn, higher capital. If this loop is left unchecked, capital would grow exponentially over time. However, as will soon be evident from the model equations, the momentum of the reinforcing loop is weakened as the balancing loops strengthen.

The first balancing loop (B_1) is captured in Table 3.5. This is a familiar depreciation loop already encountered in the previous economic model. Given the wear and tear on equipment, it will have a finite life span, and the negative feedback loop models the depreciation effect on capital.

As capital increases, so too does the cost of capital, and this in turn will reduce profits, which is shown in loop (B_2). Reduction in profits lead to lower investment levels, and hence lower capital. Table 3.6 shows the causal links that have this balancing effect on the accumulation of capital.

Finally, two more balancing loops (B_3 and B_4) combine to impact the growth potential of capital. The logic of these loops is intuitive. More capital leads to more extraction, which depletes the resource. With a lower resource, extraction efficiency declines, which lowers the extraction rate further. This leads to reduced revenue and profits, which negatively impacts capital funds. Reduced capital investment leads to a reduction in capital, therefore the direction of change for the capital stock has reversed after one iteration through the loop structure (Table 3.7).

The model equations are now presented, starting with the representation of the capital stock (3.22), which has an initial value of 5. This stock accumulates the net difference of investments and depreciation (3.23). The depreciation rate is constant at 5 % (3.24).

Table 3.7 Negative feedback loop (B_3 and B_4)—resource depletion

↑	Capital	→	Extraction	↑
↑	Extraction	→	Resource	↓
↓	Resource	→	Extraction efficiency Per unit capital	↓
↓	Extraction efficiency per unit capital	→	Extraction	↓
↓	Extraction	→	Total revenue	↓
↓	Total revenue	→	Profits	↓
↓	Profits	→	Capital funds	↓
↓	Capital funds	→	Maximum investment	↓
↓	Maximum investment	→	Investment	↓
↓	Investment	→	Capital	↓

$$Capital = INTEGRAL\ (Investments - Depreciation, 5) \tag{3.22}$$

$$Depreciation = Capital \times Depreciation\ Rate \tag{3.23}$$

$$Depreciation\ Rate = 0.05 \tag{3.24}$$

Desired investment represents the target investment rate for capital, in order to stimulate growth. It is modeled as a fixed proportion of the capital stock (3.25), where the initial goal is 7 % (3.26), and as this is greater that the depreciation rate of 5 %, the capital stock should initially grow at an exponential rate.

$$Desired\ Investment = Desired\ Growth\ Fraction \times Capital \tag{3.25}$$

$$Desired\ Growth\ Fraction = 0.07 \tag{3.26}$$

However, the non-renewable resource will ultimately limit this growth, and the stock and flow model is designed to capture this interplay. The integral equation for the resource (3.27) has an initial value of 1000, and a single outflow, which is the extraction rate (3.28). This extraction rate depends on the amount of available capital, which is multiplied by the extraction efficiency per unit of capital (3.29).

$$Resource = INTEGRAL\ (-Extraction, 1000) \tag{3.27}$$

$$Extraction = Capital \times Extraction\ Efficiency\ Per\ Unit\ Capital \tag{3.28}$$

$$\begin{gathered} Extraction\ Efficiency\ Per\ Unit\ Capital = GRAPH\ (Resource) \\ (0,0),(100,0.25),(200,0.45),(300,0.63),(400,0.75),(500,0.85), \\ (600,0.92),(700,0.96),(800,0.98),(900,0.99),(1000,1.0) \end{gathered} \tag{3.29}$$

The extraction efficiency Eq. (3.29) captures a vital relationship between the resource level and the extraction efficiency. From a technical viewpoint, it is a good example of how a stock can be used to influence a flow. This is similar to the effect

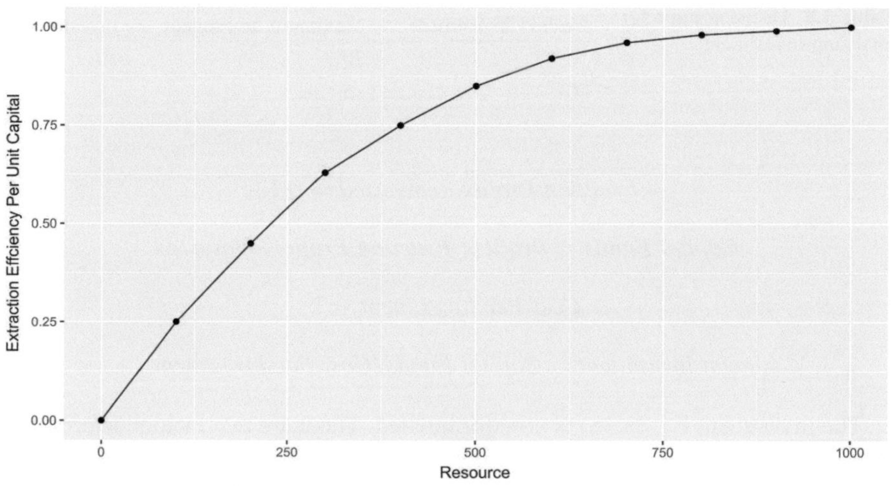

Fig. 3.8 Relationship between resource level and extraction efficiency

equation formulation discussed earlier in the chapter, as the efficiency value ranges from 1 to 0, where a value of zero will "switch off" the flow, and no further resources will be extracted, causing revenues to drop to zero. This non-linear relationship is plotted in Fig. 3.8. It shows a maximum efficiency when the resource is at its maximum value of 1000. Once the resource declines, so to does the efficiency. Initially the rate of decline is small and gradual, but once it passes the half-way mark, the efficiency drops sharply, thus impacting the outflow for the extraction process. Again, this models the scenario whereby the capability of capital extraction reduces as the oil reserves diminish.

Once the rate for extraction is calculated, the revenue and investment section of the model can be completed. The total revenue (3.30) is the amount extracted times revenue per unit extracted (3.31). The capital costs (3.32), with an arbitrary constant of 10 % used, are then deducted from the revenue to generate a value for profits (3.33).

$$\mathit{Total\ Revenue} = \mathit{Revenue\ Per\ Unit\ Extracted} \times \mathit{Extraction} \tag{3.30}$$

$$\mathit{Revenue\ Per\ Unit\ Extracted} = 3 \tag{3.31}$$

$$\mathit{Capital\ Costs} = \mathit{Capital} \times 0.10 \tag{3.32}$$

$$\mathit{Profit} = \mathit{Total\ Revenue} - \mathit{Capital\ Costs} \tag{3.33}$$

A fixed percentage of profits (3.34) are available as capital funds (3.35). The cost per unit of investment (3.36) then determines the maximum investment in capital possible (3.37).

Table 3.8 Decision logic for finalizing investment

Desired investment	Maximum investment	Investment
10	20	10
10	5	5

$$\text{Fraction Profits Reinvested} = 0.12 \tag{3.34}$$

$$\text{Capital Funds} = \text{Profit} \times \text{Fraction Profits Reinvested} \tag{3.35}$$

$$\text{Cost Per Investment} = 2 \tag{3.36}$$

$$\text{Maximum Investment} = \text{Capital Funds}/\text{Cost Per Investment} \tag{3.37}$$

The investment Eq. (3.38) is now formulated. There are two factors determining this. First, there is the desired level of investment (3.25) that is required to maintain the growth target. In a world without limits, this value would always be used in the model, and if it was, the capital stock would rise exponentially (once the growth rate exceeded the depreciation rate).

However, depending on the resources extracted and the available funding, there is the maximum possible investment that can be made (3.27), and this is the "reality check" for the system. Table 3.8 captures the required decision logic for investment. It follows the rule that the company does not invest more than its target, and that it cannot invest more than the maximum possible investment value.

In system dynamics, the conventional way to represent this type of decision between what is desired, and what can be achieved subject to constraints, is to utilize the *MIN* function, and this final equation is specified in (3.38).

$$\text{Investment} = \text{MIN}\,(\text{Desired Investment}, \text{Maximum Investment}) \tag{3.38}$$

The R model is now presented, and initially the time vectors, stocks and auxiliaries (constants) are defined. An interesting feature of this implementation is the way in which non-linear functions can be conveniently represented in R.

```
library(deSolve)

START<-0; FINISH<-200; STEP<-0.25
simtime <- seq(START, FINISH, by=STEP)
stocks  <- c(sCapital=5, sResource=1000)
auxs    <- c(aDesired.Growth       = 0.07,
             aDepreciation         = 0.05,
             aCost.Per.Investment  = 2.00,
             aFraction.Reinvested  = 0.12,
             aRevenue.Per.Unit     = 3.00)
```

Next, the nonlinear relationship between resource and extraction efficiency (3.29) is defined, and for this, R's **approxfun()** interpolation function is used. This accepts a set of x (input) and y (output) vectors. The interpolation method ("linear"), and parameters indicate the values to be returned when x is less than the minimum (yleft), and when x is greater than the maximum (yright). Two vectors are created, one for the x-axis value (**x.Resource**), and the other for the corresponding y-axis values (**y.Efficiency**). The **approxfun** function takes these vectors and creates a function that interpolates an individual resource value to its corresponding efficiency value.

```
x.Resource<-seq(0,1000,by=100)
y.Efficiency<-c(0,0.25,0.45,0.63,0.75,0.85,0.92,
                0.96,0.98,0.99,1.0)

func.Efficiency<-approxfun(x=x.Resource,
                           y=y.Efficiency,
                           method="linear",
                           yleft=0, yright=1.0)
```

The function **func.Efficiency()** implements Eq. (3.29). This can also be tested in advance for the range of values, and also for extreme cases, where the input value is outside the expected range. The following console output confirms the new function's behavior.

```
> func.Efficiency(-1)
[1] 0

> func.Efficiency(500)
[1] 0.85

> func.Efficiency(1000)
[1] 1
```

The model is now defined, where all the equations are implemented in the correct order.

```
model<-function(time, stocks, auxs){
  with(as.list(c(stocks,auxs)),
  {
    aExtr.Efficiency    <-func.Efficiency(sResource)

    fExtraction         <- aExtr.Efficiency * sCapital

    aTotal.Revenue      <- aRevenue.Per.Unit * fExtraction
    aCapital.Costs      <- sCapital * 0.10
    aProfit             <- aTotal.Revenue - aCapital.Costs
    aCapital.Funds      <- aFraction.Reinvested * aProfit
    aMaximum.Investment <- aCapital.Funds/
                           aCost.Per.Investment

    aDesired.Investment <-sCapital * aDesired.Growth

    fInvestment         <- min(aMaximum.Investment,
                               aDesired.Investment)
    fDepreciation       <- sCapital * aDepreciation

    dS_dt               <- fInvestment - fDepreciation
    dR_dt               <- -fExtraction

    return (list(c(dS_dt, dR_dt),
                 DesiredInvestment=aDesired.Investment,
                 MaximumInvestment=aMaximum.Investment,
                 Investment=fInvestment,
                 Depreciation=fDepreciation,
                 Extraction=fExtraction))
    })
}
```

The **ode** function is called, passing in the required arguments and the result is returned as a data frame.

```
o<-data.frame(ode(y=stocks, times=simtime, func = model,
                  parms=auxs, method="euler"))
```

The plots in Fig. 3.9 are now examined, in order to explore the interplay between the capital and resource stocks.

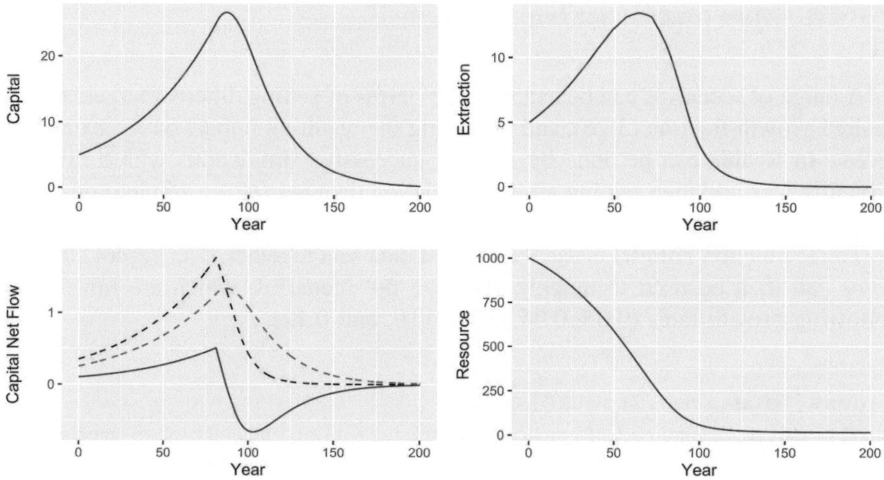

Fig. 3.9 Simulation output showing stocks and flows

- The capital initially increases exponentially, as there are sufficient resources available to ensure that growth fraction remains at the desired level, which is 7 % per annum.
- The increase in capital drives a corresponding increase in extraction, which in turn reduces the resource stock.
- Declining resources impact the capital net flow. Initially, the investment (inflow) dominates the depreciation (outflow). A critical point can be observed on the capital net flow graph, where the black line (investments) initially exceeds the red line (depreciation), and this continues until a crossover point after about 87 years, and from there on the capital falls, although the level of capital is sufficient to keep extracting the resource.

The simulation data set can be queried for exact information on peak values for capital and extraction. Using R, and the function **which.max()**, the time when capital is at its maximum can be calculated as follows.

```
> o[which.max(o$sCapital),"time"]
[1] 87
```

An important and widely used measure is when the peak of the resource extraction is reached, and this is calculated by finding the index of the maximum value of the extraction flow.

```
> o[which.max(o$Extraction),"time"]
[1] 64.25
```

A range of scenarios can be examined, in terms of setting different targets for the desired growth fraction (3.26), and observing the resulting impact on the extraction (3.28). In R, this can be done by running successive simulations with a different growth value, and then joining all the simulation data (*o*1, *o*2, ... *o*5) into one large data frame.

The R function **rbind()** is used to append data sets together. The **ggplot** attribute *color* can then be used to quickly visualize the scenarios, which are run for the following growth rates (0.05, 0.06, 0.07, 0.10, and 0.12).

```
auxs["aDesired.Growth"]<-0.05
o1<-data.frame(ode(y=stocks, times=simtime, func = model,
                   parms=auxs, method="euler"))
o1$GR<-"GR=5%"
base<-o1

auxs["aDesired.Growth"]<-0.06
o2<-data.frame(ode(y=stocks, times=simtime, func = model,
                      parms=auxs, method="euler"))
o2$GR<-"GR=6%"
base<-rbind(base,o2)

auxs["aDesired.Growth"]<-0.07
o3<-data.frame(ode(y=stocks, times=simtime, func = model,
                      parms=auxs, method="euler"))
o3$GR<-"GR=7%"
base<-rbind(base,o3)

auxs["aDesired.Growth"]<-0.10
o4<-data.frame(ode(y=stocks, times=simtime, func = model,
                    parms=auxs, method="euler"))
o4$GR<-"GR=10%"
base<-rbind(base,o4)
auxs["aDesired.Growth"]<-0.12
o5<-data.frame(ode(y=stocks, times=simtime, func = model,
                    parms=auxs, method="euler"))

o5$GR<-"GR=12%"
base<-rbind(base,o5)
ggplot(data=base,aes(x=time,y=Extraction,color=base$GR))+
       geom_line()+ xlab("Year")+ylab("Extraction Rate")+
       theme(legend.position="bottom")+
       guides(color=guide_legend(title=NULL))
```

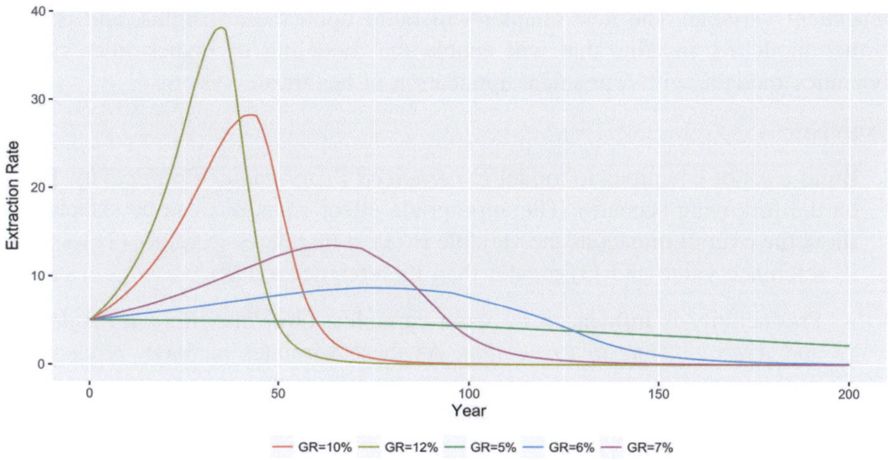

Fig. 3.10 Impact of desired growth rates on resource extraction

Table 3.9 Comparing extraction rate attributes based on different desired growth rates

Desired growth rate (%)	5	6	7	10	12
Peak value	5.0	8.70	13.49	28.24	38.15
Peak time	0	71.875	64.125	42.5	34.625

The range of extraction rates are visualized in Fig. 3.10. As expected, higher desired growth rates lead to steeper extraction rates, and quicker depletion of the resource. It confirms the view of Meadows (2008) that "a quantity growing exponentially toward a constraint or a limit reaches that limit in a surprising short time", and that "the higher and faster you grow, the farther and faster you fall."

Data on the individual simulation runs is summarized in Table 3.9. This shows the impact of an increasing desired growth rate on the peak extraction value and the peak time of extraction. Higher growth rates lead to higher peak values, but also lead to earlier peak times.

Summary

This chapter presented limits to growth models, where the availability of a resource impacts a system's growth potential. These models are relevant in many constraint-based problems, including business, healthcare and resource extraction industries. This chapter also demonstrated an important system dynamics technique which allows a number of independent variables to influence the value of a dependent variable. The next chapter will build upon these insights, and present further modeling insights that will enable the modeling of higher-order system dynamics models, with a practical application in healthcare systems.

Exercises

1. Build a set of equations to model *Experienced Programmer Productivity,* based on the following scenario. The appropriate effect equations can be sketched to show the overall impact as the variable is (1) at its reference value, (2) less than its reference value and (3) greater than its reference value.
   - Productivity is influenced by three variables: *Overtime*, *Rookie Proportion* and *Average Time to Promotion.* As these variables increase, productivity declines.
   - The reference value for *Experienced Programmer Productivity* is 200 lines of code (LOC)/Day.
   - The reference value for overtime is 5 h per week.
   - The reference rookie proportion is 20 %.
   - The average time for promotion is 24 months.
2. Find an analytical solution to the following representation of the logistic growth model, where P is the population, r is the growth rate, and K is the carrying capacity.

$$\frac{dP}{dt} = rP\left(1 - \frac{P}{K}\right)$$

3. Based on the non-renewable stock model, and assuming a capital growth rate of 10 %, run two additional scenarios whereby the resource is doubled and quadrupled. What impact does these additional scenarios have on the time of peak extraction?

References

Bacaër N (2011) Verhulst and the logistic equation (1838). In: A short history of mathematical population dynamics. Springer London, pp 35–39
Meadows DH (2008) Thinking in systems: a primer. Chelsea Green Publishing
Page S (2015) A model of growth. Supporting material for Coursera Model Thinking MOOC Course. https://www.coursera.org/course/modelthinking. Accessed 30 June 2015
Richardson GP (1991) Feedback thought in social science and systems theory. Pegasus Communications, Inc., Chicago
Solow RM (1956) A contribution to the theory of economic growth. Quart J Econ 70:65–94
Sterman JD (2000) Business dynamics: systems thinking and modeling for a complex world. Irwin/McGraw-Hill, Boston

Verhulst P-F (1845) "Recherches mathématiques sur la loi d'accroissement de la population." Nouv. mém. de l'Academie Royale des Sci. et Belles-Lettres de Bruxelles 18:1–41
Verhulst P-F (1847) "Deuxième mémoire sur la loi d'accroissement de la population." Mém. de l'Academie Royale des Sci., des Lettres et des Beaux-Arts de Belgique 20:1–32

Chapter 4
Higher Order Models

Important situations in management, economics, medicine and social behavior often lose reality if simplified to less than fifth-order nonlinear dynamic systems.
Often the model representation must be twentieth-order or higher.

Jay W. Forrester (1987).

Abstract This chapter presents a higher order model, which has a greater number of stocks and feedbacks than those presented in earlier chapters. This is an important perspective, as real-world system dynamics models tend to have a significant number of stocks. To aid understanding, higher order models are often sub-divided into distinct *sectors*, where each sector contains a recognizable sub-system. This higher order model represents a primary health care system that models an aging demographic, the supply of general practitioners, and the annual demand the population places onto the primary care system. Before presenting this model two important modeling constructs are described. These are *delays*, which allow modelers to simulate time lags, and the *stock management structure*, which provides a structure to simulate how decision makers regulate the stock levels.

Keywords Delays · Stock management structure · Sectors · Health system · Demographics

Delays

Delays are a feature of many social and business systems, and stock and flow structures can be used to model delays. Example of delays include:

- A software company may have an innovative idea for a new product, but building a software system takes time. Requirements must be gathered from prospective users, a design needs to be architected, and the system must be coded and tested.

J. Duggan, *System Dynamics Modeling with R*, Lecture Notes in Social Networks,
DOI 10.1007/978-3-319-34043-2_4

- Career progression contains delays. Employees work in defined roles, and as time progresses they move on to more senior posts.
- Students enter a university in first year, and progress through successive years until they graduate. This process takes a number of years.
- For contagious diseases, people are infectious for a certain time period, and during this time they can transmit the disease. After a time delay, they recover, and are no longer infectious.

Forrester (1961) defines a delay as a "conversion process that accepts a given inflow rate and delivers a resulting outflow rate." Delays *always* contain a stock, as the stock structure captures the accumulation of material in transit. In system dynamics, the simplest delay structure is known as a *first order exponential delay*, because the delay distribution is modeled using just one stock. For example, in a software development environment, the stock and flow structure for fixing bugs (i.e. rework) is shown in Fig. 4.1.

The stock models the accumulation of bugs that need to be fixed (4.1), where the errors found (4.2) is the inflow, and the outflow is errors fixed (4.3). In this scenario, the inflow starts at 100 and then is stepped up by 50 after 1 time unit, thereby modeling the discovery of additional bugs. For a first-order delay, the outflow is the current stock divided by the average fix time, where the average fix time is an arbitrary constant (4.4). This average fix time can also be inverted and conveniently expressed as a proportion (4.5), and this can also be a useful way to formulate, and communicate, a delay.

$$Rework = INTEGRAL\ (Errors\ Found - Errors\ Fixed, 600) \tag{4.1}$$

$$Errors\ Founds = 100 + STEP\ (50, 1) \tag{4.2}$$

$$Errors\ Fixed = Rework/Average\ Fix\ Time \tag{4.3}$$

$$Average\ Fix\ Time = 6 \tag{4.4}$$

$$Average\ Fix\ Proportion = 1/Average\ Fix\ Time \tag{4.5}$$

Higher-order delays are constructed by cascading a number of first order delays together, causing the material in transit to progress through a series of intermediary

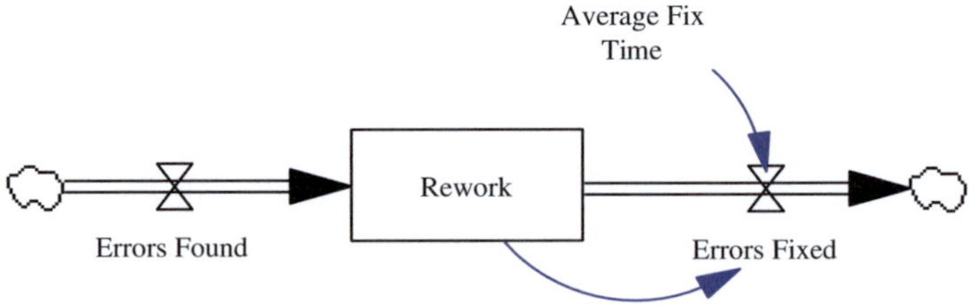

Fig. 4.1 A first order material delay

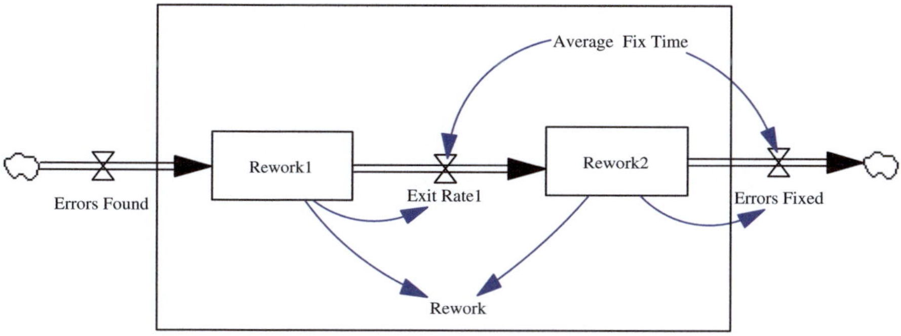

Fig. 4.2 A second order material delay

stocks. The overall delay time is averaged out equally across all the stock outflows. These stocks do not have a physical equivalent in a real system, they are solely used to model the appropriate delay response. For example, a *second-order exponential delay* involves linking two first order delays together, and Fig. 4.2 shows how the software example can be represented as a second order delay.

While the first-order delay modeled material in transit in a single stock with a delay of six time units, a second order delay has two sequential stocks (4.6 and 4.7). The outflow from the first stock is the inflow to the second stock, and, initially, each stock contains half the contents of the overall delay. Note that the model will start in equilibrium, as all the flows will still equal 100, given that the time delay for the individual stocks is the overall time delay divided by two. The total amount of material in transit is simply the sum of these two stocks (4.10).

$$Rework1 = INTEGRAL\ (Errors\ Found - Exit\ Rate1, 300) \tag{4.6}$$

$$Rework2 = INTEGRAL\ (Exit\ Rate1 - Errors\ Fixed, 300) \tag{4.7}$$

$$Exit\ Rate1 = Rework1/(Average\ Fix\ Time/2) \tag{4.8}$$

$$Errors\ Fixed = Rework2/(Average\ Fix\ Time/2) \tag{4.9}$$

$$Rework = Rework1 + Rework2 \tag{4.10}$$

There are two characteristics of delays that are of interest (Forrester 1961). First, is the *delay duration*, which is the average time material spends in the delay. This value also determines the stock's value when the system is in equilibrium, which occurs when the inflow equals the outflow. For equilibrium, the quantity of material in transit (i.e. the stock) is the flow rate multiplied by the average delay. For example, the first order rework model starts in equilibrium, as the inflow (100) equals the outflow, and the material in transit is six times this value (600). This steady-state relationship is also known as *Little's Law* (Cachon and Terwiesch 2009), which is a

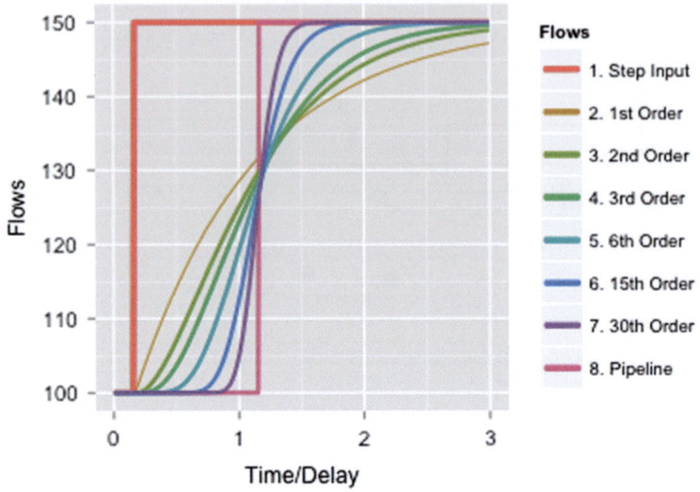

Fig. 4.3 Range of transient responses for system dynamics delays

very useful heuristic that can be used to determine the value of the stock in equilibrium, and is summarized in Eq. 4.11.

$$\textit{Material in Transit} = \textit{Average Flow Rate} \times \textit{Average Flow Time} \quad (4.11)$$

The second delay characteristic is the *transient response* of the delay, as this shows how the behavior of the *outflow* relates to the behavior of the *inflow*. Delay structures have different transient responses, and the most suitable transient response should be selected based on available data. Consider the transient responses of seven different delay structures, and how they respond to the step change of 50 units from an initial equilibrium value of 100, as shown in Fig. 4.3. The x-axis contains the ratio of the simulation time to the delay duration.

The first output is the first-order exponential delay, where the response is initially immediate, and this structure can be an appropriate model for certain processes. For example, Coyle (1996) suggests how this is could be a suitable model for a bus company which takes on qualified drivers, as some will be productive quite quickly, while others will take longer to learn the routes. The higher the delay order (for example, 15th and 30th order in our example), the more the output response takes on the shape of the input step change, and this can be seen from the figure, as the higher order delays move towards the pattern of the delay input.

An infinite-order delay is known as a *pipeline delay*, where the output exactly matches the input after a fixed duration. An example of a pipeline day could be a distribution process, where 1000 units leave an arrival point at the same time, and are delivered together after a fixed duration. Most system dynamics tools can accommodate pipeline delay structures. However, in social systems, where there is significant variability in delay output, combinations of first, second and third order delays can model the required dynamics. In this text, the models presented make use of the first order delay, although in the system dynamics literature, higher order models are also commonly used. A number of system dynamics textbooks contain extensive treatments of delays and their dynamics, and the interested reader is referred is the works of Forrester (1961) and Sterman (2000).

The Stock Management Structure

In many systems, a challenge is to manage stocks to ensure they remain at desired levels. For example, consider the following scenarios.

- Retailers must maintain inventories at defined levels to ensure a reliable flow of goods to customers.
- Universities need to maintain student levels at desired numbers.
- Companies have to regulate their employee levels so as to ensure that goods and services can be produced.

These are all examples of stock management challenges, and system dynamics provides a valuable generic structure for modeling these scenarios, in order to allow decision makers to explore the consequences of management policies.

A stock management structure, focusing on managing employee levels, is shown in Fig. 4.4. Employees are increased by new hires, and depleted by employees

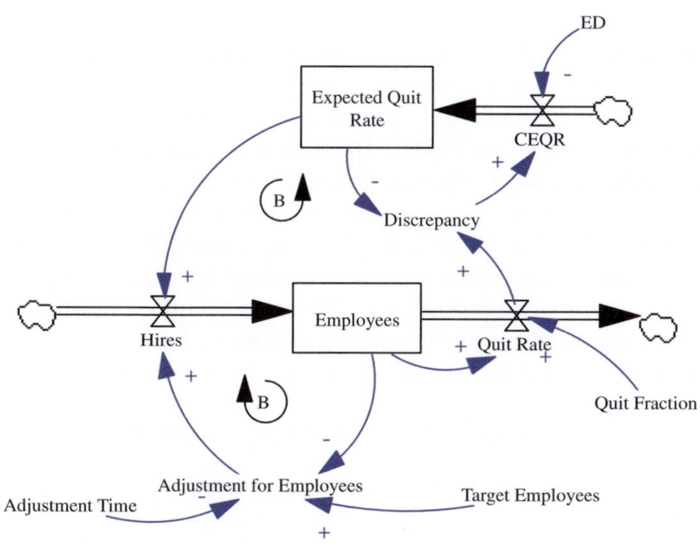

Fig. 4.4 The stock management structure

leaving. The stock management heuristic seeks to maintain the employee stock at the desired level, and has two components.

- First, in order to manage the stock, an expectation of employee losses is required, as these are future outflows will need to be replaced. For example, if a company has an average churn rate of 12 % employees per year, then it should expect to lose 1 % of its employees per month. In order to maintain services at current levels, this will require monthly recruitment to cover these losses.
- Second, in addition to replacing expected losses, managing a stock also requires maintaining the stock at desired levels. This desired level can vary over time. For example, seasonal demand for a service industry would require additional staff to maintain output levels, and therefore the desired level of staff would rise. The stock management rule needs to account for adjusting the stock towards it desired level.

The goal of the stock management heuristic is to formulate the stock's inflow rate (in this case the number of hires). The main stock of employees is shown in Eq. (4.12) as the integral of inflows (hires) minus outflows (the quit rate), where the stock's initial value is 100 employees. The quit rate (4.13) is defined as the number of employees times the quit fraction (4.14).

$$Employees = INTEGRAL\left(Hires - Quit\ Rate, 100\right) \tag{4.12}$$

$$Quit\ Rate = Employees \times Quit\ Fraction \tag{4.13}$$

$$Quit\ Fraction = 0.10 \tag{4.14}$$

As simulation mimics the way a system operates, the stock management structure must capture how the decision maker uses information and transforms this into action. In order to model the decision maker's expectation of losses, a new variable must be added to the model. This is a *first-order information delay*, and uses a stock (4.18) to model expected losses. This variable can be thought of as a "mental" stock, namely it represents the decision maker's perception as to the variable's future value. This stock is simply an adjustment based on the discrepancy between actual value and the expectation (4.15 and 4.16), factored by a smoothing constant (4.17). What is interesting about this stock and flow structure is that it is mathematically equivalent to an exponential smoothing forecasting process, which is a widely heuristic used for planning purposes (Forrester 1961).

$$Discrepancy = Quit\ Rate - Expected\ Quit\ Rate \tag{4.15}$$

$$CEQR = Discrepancy/ED \tag{4.16}$$

$$ED = 2 \tag{4.17}$$

$$Expected\ Quit\ Rate = INTEGRAL\ (CEQR, 10) \tag{4.18}$$

The second component of the stock management decision heuristic is to adjust the stock towards its desired value. This is a goal-seeking decision rule that calculates the gap between desired employees and the current number (4.19), and modifies this using an adjustment constant (4.21). The value of this constant determines how quickly the system reaches its goal. The target number of employees (4.20) starts at an equilibrium value of 100, and the steps up after 6 time units. The step change triggers a response from the stock management structure, where the behavior is to readjust the flows until the system finds a new equilibrium.

$$Adjustment\ for\ Employees = (Target\ Employees - Employees)/AT \tag{4.19}$$

$$Target\ Employees = 100 + STEP\ (50, 6) \tag{4.20}$$

$$AT = 4 \tag{4.21}$$

The final element of the stock management structure is the hire rate, and this is a summation of expected stock losses, and adjustments to the stock (4.22). The *MAX* function formulation ensures that the hire rate always stays positive.

$$Hire\ Rate = MAX\ (0, Expected\ Quit\ Rate + Adjustment\ for\ Employees) \tag{4.22}$$

The behavior of the stock management structure is shown in Fig. 4.5, with the stock and target on the left, and the two components of the hire rate on the right. The system starts in equilibrium with the number of employees at its target level (100). The hire rate is 10, which covers expected losses (at 10 % of the stock). Because the system is at its target value, no adjustment is needed. The system is then nudged out of equilibrium when the target changes by 50. The response is interesting. First, the adjustment immediately responds, given that more employees are required. Also, the expected losses increase as more employees are added. By time 30, the adjustment has dropped back to zero, and the system is in equilibrium once more. At this point, expected losses have reached 15, which is 10 % of the new stock target.

The stock management structure is a negative feedback system. It takes into account the expected losses from the stock and replaces these. It also seeks to bridge any gap between the desired stock value, and the current value. In the next example, the stock management structure plays an important role in regulating the supply of general practitioners.

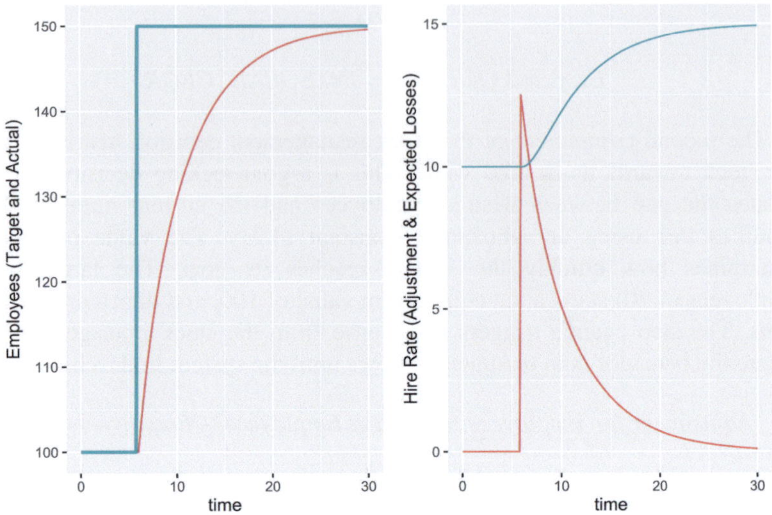

Fig. 4.5 Simulation output for the stock management structure

Health Care Model

The formulations for the stock management structure, first order delays, and the effects models from Chap. 3 are now combined in a health care model. This model simulates the impact of demographic change on demands for primary care services. System dynamics has been successfully applied to complex problems in health care, as part of efforts to reduce morbidity and mortality, including: health care delivery, population health and health economics, substance abuse, infectious disease, biology and microbiology and health care products (Hirsch et al. 2013). A key benefit of system dynamics is that it accommodates a broad, system-wide, perspective that captures a wide system boundary. For example, Forrester's *market growth model,* which shows how poor decision making can impact market growth in a potentially infinite marketplace, also has a number of interdependent sectors, including customer demand, order fulfillment, sales and capacity acquisition (Forrester 1968). In a similar vein, this health care delivery model has three interacting sectors:

- A *demographic* sector, which is an aging chain that captures the dynamics of population change, across a number of age cohorts.
- A *delivery* sector, which is a demand-capacity model that captures how the primary care system responds to demand.
- A *supply* sector, which contains a stock management structure that regulates the supply of general practitioners, in proportion to the overall population.

Sectors are a convenient way to partition models, and a sector represents a cohesive set of equations for a sub-system. The model sectors are shown in

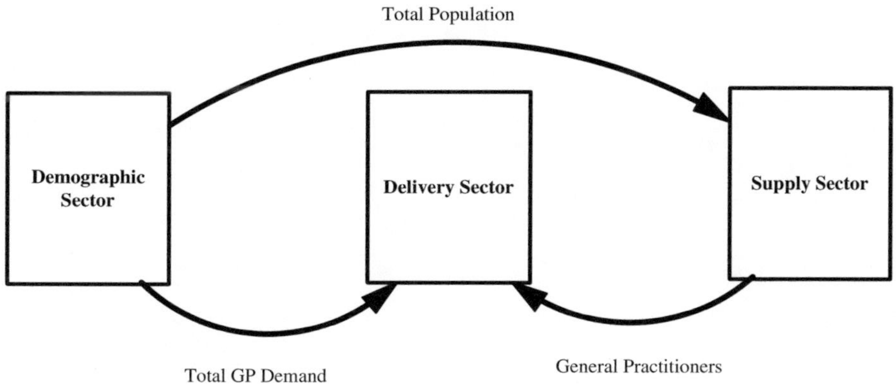

Fig. 4.6 Model sectors

Fig. 4.6, along with the key information flows between the three sectors. For example, this shows that the total general practitioner demand in the delivery sector is determined by stocks contained in the demographic sector. The three sectors are now described.

Demographic Sector

The demographic sector, shown in Fig. 4.7, is an *aging chain* structure that simulates population maturation, as well as births and deaths. The initial population size is 5 M, with the initial cohort values at (1 M, 1.5 M, 2.0 M and 0.5 M) respectively. This gives an initial dependency ratio[1] of 42.8 per hundred, and higher ratios place greater stresses on social and health services. This sector also generates the total demand for general practitioners services, based on published data on the estimated average visits per cohort per year. There are a number of assumptions behind this demographic model, including:

- The number of cohorts is simplified to four, and no distinction is made between male and female. Also, there is no immigration or emigration in the model, and all the removals are from the oldest cohort.
- First order delays are used to model cohort progression, where the average delay time is 15 years for the first cohort, and 25 years for the other cohorts.
- Births are based on a fixed proportion of the total population. The birth and death rates are exogenous, which is a limitation of this initial model.

[1]The dependency ratio is a standard economic measure that captures the proportion of non-working ($P_{0-14} + P_{65+}$) to working ($P_{15-39} + P_{40-64}$) population.

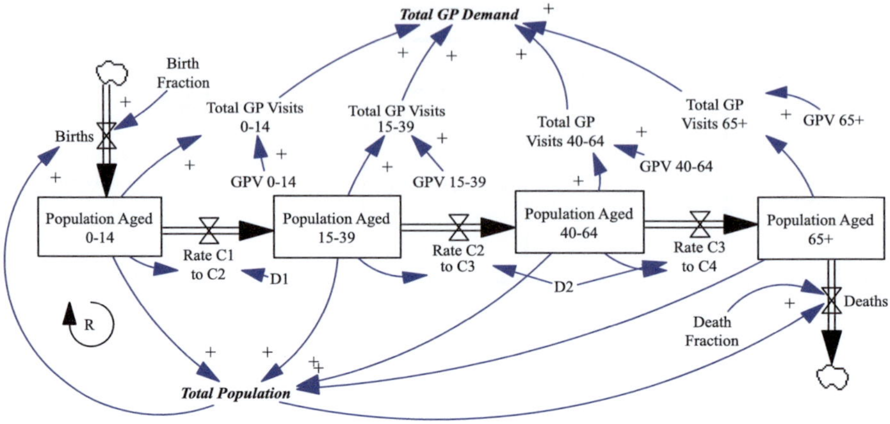

Fig. 4.7 The demographic sector

The model comprises four sequential cohorts (4.23–4.26), where each stock has one inflow and one outflow. Births add to *Population Aged 0–14* (P_{0-14}), and the first progression rate fills *Population Aged 14–39* (P_{15-39}). A similar structure is used for the remaining stocks *Population Aged 40–64* (P_{40-64}) and *Population Aged 65+* (P_{65+}). The total population is the sum of these four stocks (4.27).

$$P_{0-14} = INTEGRAL\ (Births - Rate\ C_1\ to\ C_2, 1.0\,M) \tag{4.23}$$

$$P_{15-39} = INTEGRAL\ (Rate\ C_1\ to\ C_2 - Rate\ C_2\ to\ C_3, 1.5\,M) \tag{4.24}$$

$$P_{40-64} = INTEGRAL\ (Rate\ C_2\ to\ C_3 - Rate\ C_3\ to\ C_4, 2.0\,M) \tag{4.25}$$

$$P_{65+} = INTEGRAL\ (Rate\ C_3\ to\ C_4 - Deaths, 0.5\,M) \tag{4.26}$$

$$Total\ Population = P_{0-14} + P_{15-39} + P_{40-64} + P_{65+} \tag{4.27}$$

The flows are captured in Eqs. (4.28–4.32). For simplicity, births are calculated on the aggregate population, and are driven by a positive feedback loop, which is dominant once the birth fraction exceeds the death fraction. A more detailed birth model would formulate births on the female cohort size of child bearing age, along with an estimate of overall fertility. Progression rates are first order delays, and the deaths are based on the overall death fraction applied to the total population, and removed from the oldest cohort. (Note: this assumption is made to simplify the number of outflows from the model.)

$$Births = Total\ Population \times Birth\ Fraction \tag{4.28}$$

$$Rate\ C_1\ to\ C_2 = P_{0-14}/D_1 \tag{4.29}$$

$$Rate\ C_2\ to\ C_3 = P_{15-39}/D_2 \tag{4.30}$$

$$Rate\ C_3\ to\ C_4 = P_{40-64}/D_2 \tag{4.31}$$

$$Deaths = Total\ Population \times Death\ Fraction \tag{4.32}$$

The relevant model constants include the birth fraction (4.33), death fraction (4.34), and time delays (4.35 and 4.36).

$$Birth\ Fraction = 20/1000 \tag{4.33}$$

$$Death\ Fraction = 7/1000 \tag{4.34}$$

$$D_1 = 15 \tag{4.35}$$

$$D_2 = 25 \tag{4.36}$$

Given that the primary function of the demographic model is to generate realistic demand patterns for the primary care sector, an estimate for annual general practitioner visits (GPV) is used based on available data (Lyons and Duggan 2015). This shows a gradual increase in annual visit rates for the first three cohorts (4.37–4.39), followed by a sharp increase for the elderly cohort (4.40).

$$GPV_{0-14} = 3 \tag{4.37}$$

$$GPV_{15-39} = 4 \tag{4.38}$$

$$GPV_{40-14} = 5 \tag{4.39}$$

$$GPV_{65+} = 10 \tag{4.40}$$

Based on these visiting rates, and the population size for each cohort, the total general practitioner visits (TGVP) for each group is calculated (4.41–4.44), and following that the total aggregate demand for services is calculated (4.45).

$$TGPV_{0-14} = P_{0-14} \times GPV_{0-14} \tag{4.41}$$

$$TGPV_{15-39} = P_{15-39} \times GPV_{15-39} \tag{4.42}$$

$$TGPV_{40-64} = P_{40-64} \times GPV_{40-64} \tag{4.43}$$

$$TGPV_{65+} = P_{65+} \times GPV_{65+} \tag{4.44}$$

$$Total\ GP\ Demand = TGPV_{0-14} + TGPV_{15-39} + TGPV_{40-64} + TGPV_{65+} \tag{4.45}$$

This final value for total general practitioner demand is the main output from this sector, and this value is used to determine patient demand in the delivery sector model.

Delivery Sector

The delivery sector is informed by the service capacity model described by Oliva (1996, 2001), and Sterman (2000). It provides a convenient structure (Fig. 4.8) to model resource-constrained systems. An elaborate form of a first order delay, it takes into consideration the overall demand, in terms of patient visits, and the completion rate as patient demand is fulfilled. The model contains variables that model capacity, which include the length of the work year, the average daily productivity of GPs, and the number of available general practitioners.

The model features balancing loops to cater for system responses to increases in work pressure. These loops model the actions available to GPs when demand exceeds capacity. The policy responses are in loops B_1, where extra days are worked to cope with increasing demand, and B_2 where the general practitioner productivity is increased, resulting in more patient visits per day. From a modeling

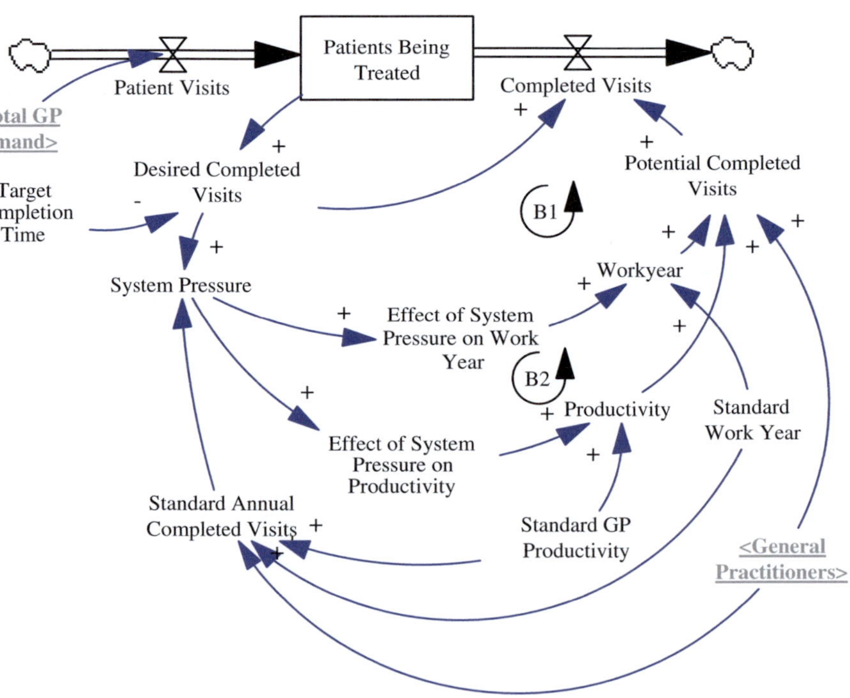

Fig. 4.8 The delivery sector

perspective, the activation of these separate policy loops can be controlled, and this process is illustrated later in the chapter. The stock (4.46) is Patients Being Treated (PBT), and this has an inflow (4.47) which is determined by the demographic sector (4.45).

$$PBT = INTEGRAL\,(Patient\ Visits - Completed\ Vists, 24\,M) \tag{4.46}$$

$$Patient\ Visits = Total\ GP\ Demand \tag{4.47}$$

Next, the desired number of completed visits is formulated (4.48). This value is the number of patient visits that would be completed if there were no resource constraints operating in the system. In effect, this value represents the number of patients who need to be treated in any given year.

$$Desired\ Completed\ Visits = PBT/Target\ Completion\ Time \tag{4.48}$$

$$Target\ Completion\ Time = 1 \tag{4.49}$$

However, health care systems have limits, and the available capacity can be calculated in terms of the number of standard annual completed visits (4.50) that are feasible. This is the product of the number of GPs (4.60), the standard work year (4.51) and the standard GP productivity, in terms of visits per day (4.52). In this example, the product of these values would give (4000 × 250 × 24) = 24 M people/year as the total system capacity.

$$\begin{aligned} Standard\ Annual\ Completed\ Visits = {} & General\ Practitioners \times Standard\ Workyear \\ & \times Standard\ GP\ Productivity \end{aligned} \tag{4.50}$$

$$Standard\ Workyear = 250 \tag{4.51}$$

$$Standard\ GP\ Productivity = 24 \tag{4.52}$$

The question then arises as to whether the system's available capacity can cope with the demographic demands. The variable system pressure (4.53) is a useful ratio that reflects how well capacity can meet demand. If this value exceeds 1, it signals that there is insufficient capacity to meet demands, and therefore the queues for treatment will lengthen, unless actions are taken.

$$\begin{aligned} System\ Pressure = {} & Desired\ Completed\ Visits \\ & /Standard\ Annual\ Competed\ Visits \end{aligned} \tag{4.53}$$

System pressure is an important information cue that informs policy responses. There are two potential responses that general practitioners can take when demand exceeds capacity.

The first response is to extend the work year so that additional visits can be scheduled, and this policy is captured using an effect variable (4.54). This relationship, which is based on empirical data from Oliva's (1996) model of service quality delivery, indicates that as the system pressure increases beyond 1, so too does the multiplier effect on the actual work year. The effect equation also models the impact of lower demand, which leads to a reduction in the number of days worked. This is beneficial, as it models GPs reducing their availability in order to balance demand with capacity. The annual work year is the product of the effect with the standard work year, and this is captured in Eq. (4.55).

$$\begin{aligned}&\textit{Effect of System Pressure on Work Year}\\&\quad = \textit{GRAPH}\ (\textit{System Pressure})\\&\qquad (0.0, 0.75), (0.25, 0.79), (0.5, 0.84), (0.75, 0.90), (1.0, 1.0), (1.25, 1.09),\\&\qquad (1.5, 1.17), (1.75, 1.23), (2.0, 1.25), (2.25, 1.25), (2.5, 1.25)\end{aligned} \tag{4.54}$$

$$\begin{aligned}\textit{Workyear} = {} & \textit{Effect of System Pressure on Workyear}\\ & \times \textit{Standard Workyear}\end{aligned} \tag{4.55}$$

The second policy is to increase daily productivity so that a greater number of patients are seen each day. Throughout many industries, it is common to use this approach, and this feedback loop is often termed *burning the midnight oil.* A possible side-effect of this policy is the impact on quality, as shorter consultation times could lead poorer diagnosis, and so increase appointments at a later time. The productivity effect is captured in (4.56), and it is also based on empirical data from the service industry (Oliva 1996). The overall impact of schedule pressure on productivity is modeled in Eq. 4.57.

$$\begin{aligned}&\textit{Effect of System Pressure on Productivity}\\&\quad = \textit{GRAPH}\ (\textit{System Pressure})\\&\qquad (0.0, 0.62), (0.2, 0.65), (0.4, 0.84), (0.6, 0.79), (0.8, 0.89), (1.0, 1.0),\\&\qquad (1.2, 1.14), (1.4, 1.24), (1.6, 1.32), (1.8, 1.37), (2.0, 1.4)\end{aligned} \tag{4.56}$$

$$\begin{aligned}\textit{Productivity} = {} & \textit{Effect of System Pressure on Productivity}\\ & \times \textit{Standard GP Productivity}\end{aligned} \tag{4.57}$$

The final two equations in this sector determine the system capacity. The potential completed visits (4.58) is the product of GPs, productivity and work year. This provides information to formulate the outflow on the stock, and for this, the minimum of desired completed visits and potential completed visits is used (4.59).

This robust formulation ensures that the stock will never go negative, and that the outflow cannot exceed the available operational capacity.

$$PotentialCompleted\ Visits = General\ Practitioners \times Productivity \times Workyear \tag{4.58}$$

$$Completed\ Visits = MIN\ (Desired\ Completed\ Visits, Potential\ Completed\ Visits) \tag{4.59}$$

Supply Sector

The supply sector models the GP resource base in terms of recruitment into the profession, and retirement after many years of service. The model, based on the stock management structure, shown in Fig. 4.9, has a number of assumptions.

- It assumes there is a ready supply of qualified personnel ready to enter practice. For a more comprehensive model, this would have to be revisited, and a full education supply line added.
- The estimation for the desired number of GPs is a crude measure, based on a fraction of the overall population. This will ensure that the number of GPs grow as the population grows.

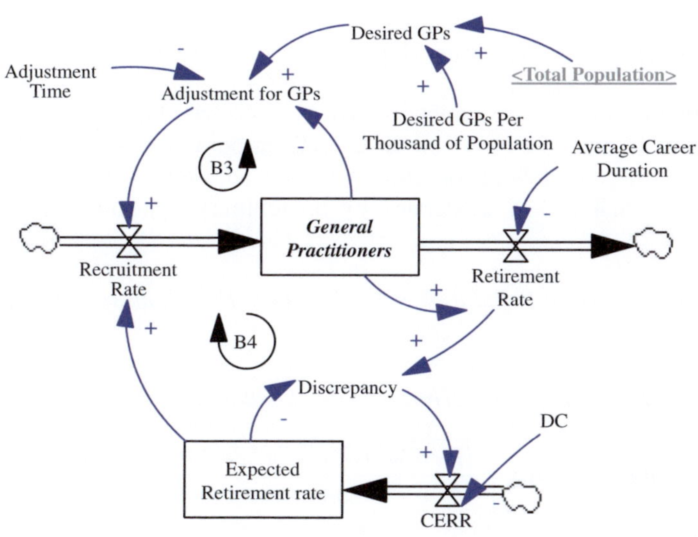

Fig. 4.9 The supply sector

- There is no distinction between the different stages in a GP professional's career. These changes, if made, could impact on the available work year, and also influence overall productivity (Fig. 4.9).

However, the model does provide a useful structure to capture the underlying resource base for the delivery sector, and so observe how the demographic changes impacts intake into the profession. As it is a stock management structure, the goal is to maintain the stock at a desired level, and also account for future losses to the stock. The general practitioners stock (4.60) has an initial value of 4000. It accumulates those recruited, and losses arise from retirements (4.61). In this model, the average career duration is 40 years (4.62).

$$General\ Practitioners = INTEGRAL\ (Recruitment\ Rate - Retirement\ Rate, 4000) \tag{4.60}$$

$$Retirement\ Rate = General\ Practitioners/Average\ Career\ Duration \tag{4.61}$$

$$Average\ Career\ Duration = 40 \tag{4.62}$$

The expected losses for the model is the expected retirement rate (4.63), which is an information delay on the retirement rate, based on the discrepancy (4.66) and delay constant (4.65).

$$Expected\ Retirement\ Rate = INTEGRAL\ (CERR, 100) \tag{4.63}$$

$$CERR = Discrepancy/DC \tag{4.64}$$

$$DC = 3 \tag{4.65}$$

$$Discrepancy = Retirement\ Rate - Expected\ Retirement\ Rate \tag{4.66}$$

The target for the desired number of GPs (4.67) is based on the total population, and is a simple measure based on an overall proportion of 0.8 per thousand of population (4.68). Following on from this, the adjustment is the gap between desired and actual (4.69), moderated by an arbitrary adjustment time constant (4.70).

$$Desired\ GPs = Total\ Population \times Desired\ GPs\ Per\ Thousand\ of\ Population \tag{4.67}$$

$$Desired\ GPs\ Per\ Thousand\ of\ Population = 0.8/1000 \tag{4.68}$$

$$Adjustment\ for\ GPs = \frac{(Desired\ GPs - General\ Practitioners)}{Adjustment\ Time} \tag{4.69}$$

$$Adjustment\ Time = 5 \tag{4.70}$$

Finally, the recruitment rate (4.71) for this stock management structure is the sum of expected retirements and the adjustment.

$$\begin{aligned} Recruitment\ Rate = MAX(0, & Expected\ Retirement\ Rate \\ & + Adjustment\ for\ GPs) \end{aligned} \tag{4.71}$$

With the three model sectors specified, an initial policy analysis can now be conducted by running the simulation model under two different scenarios.

Scenario Analysis for the Health Care Model

The power of system dynamics simulation is that it enables the exploration of specific policy issues, and for this model a single issue, of potentially many, is explored. This addresses sustainability, and the goal of the simulation is to evaluate the long-term sustainability, in terms of GP capacity, of the system. With this in mind, there are two scenarios explored.

- Scenario 1—*Business as usual*—is where the simulation is run without any policy interventions, other than the strategy to increase general practitioner numbers in direct proportion to the population size increases (4.68). For this scenario, the two policy responses to increased demand are deactivated through binary flags, and reformulated (4.72 and 4.73) using an if-else statement. When the flag is any value other than 1, the negative feedback loops B_1 and B_2 are deactivated.

$$\begin{aligned} Workyear = IF\ THEN\ ELSE\ (& Workyear\ Flag = 1, \\ & Effect\ of\ System\ Pressure\ on\ Work\ Year \\ & {}^{*}\ Standard\ Work\ Year, Standard\ Work\ Year) \end{aligned} \tag{4.72}$$

$$\begin{aligned} Productivity = IF\ THEN\ ELSE\ (& Productivity\ Flag = 1, \\ & Effect\ of\ System\ Pressure\ on\ Productivity \\ & {}^{*}\ Standard\ GP\ Productivity, \\ & Standard\ GP\ Productivity) \end{aligned} \tag{4.73}$$

- Scenario 2—*Flexible Capacity*—where in response to system pressure, addition capacity strategies in terms of (1) a longer work year and (2) increased productivity are activated, which means that the two flags are set to 1, and therefore the policy response feedback loops are activated.

Table 4.1 Calculation of initial system capacity

Standard GP productivity	24 patients/GP/day
Standard work year	250 days/year
General practitioners	4000 GPs
Standard annual capacity	24,000,000 patients/year

Table 4.2 Calculation of initial system demand

Cohort	Initial value	Average visits	Initial visits
$Population_{0-14}$	1,000,000	3	3,000,000
$Population_{15-39}$	1,500,000	4	6,000,000
$Population_{40-64}$	2,000,000	5	10,000,000
$Population_{65+}$	500,000	10	5,000,000
Total initial visits			24,000,000

The system is setup to start in equilibrium, and this can be seen through the standard capacity value, which is 24 M visits/year, and the initial demand, which is also 24 M visits/year, as calculated in Tables 4.1 and 4.2.

The simulation is run from 2014 to 2050, and results in an overall increase in population, given that the birth fraction (4.33) always exceeds the death fraction (4.34) by 13 per thousand of population. Over time, this dynamic sees the overall population grow from 5 M to 7.98 M in 2050, as shown in Fig. 4.10. Interestingly, the model shows the composition of the population changes over time, with the

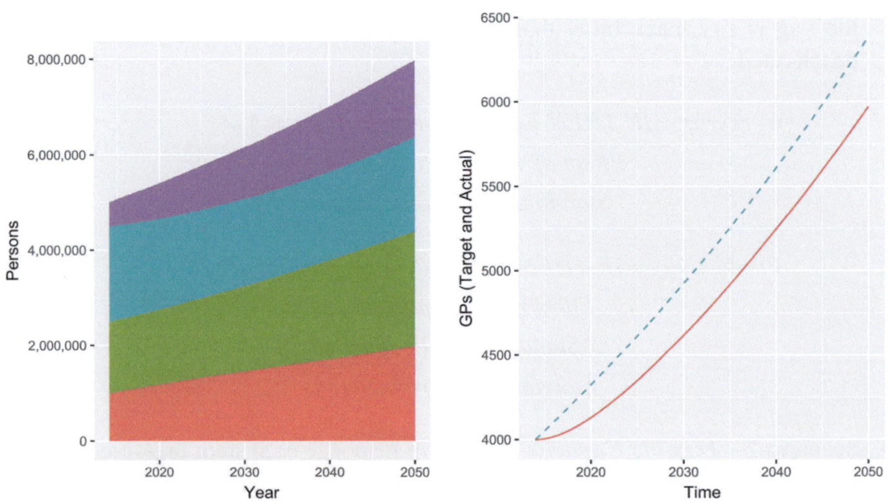

Fig. 4.10 Increasing population leads to investment in GPs

Table 4.3 Snapshots of population profile over time

Year	$Population_{0–14}$ (%)	$Population_{15–39}$ (%)	$Population_{40–64}$ (%)	$Population_{65+}$ (%)
2014	20	30	40	10
2025	23	29	32	16
2035	24	29	28	19
2050	25	30	25	20

elder age cohort increasing from 10 % of the population in 2014, to 20 % in 2050. A summary of these demographic changes are captured in Table 4.3.

While this number is based on model simplifications (e.g. all deaths subtracted from eldest cohort with the entire population only divided into four cohorts), it does capture the common demographic dynamic of our era, where the proportion of elderly is increasing over time. Figure 4.10 also shows the corresponding increase in general practitioners. The stock management structure responds to increased population size, whereby the target number of GPs is a fixed proportion of the overall population (4.67). The lag between target and actual is due to the adjustment time (4.70).

This underlying momentum of an increasing aging population has a significant impact on the delivery system, as can be inferred from the base data on average visits per cohort type. These impacts can be observed by running the two scenarios, *business as usual* (1) and *flexible capacity* (2), and highlighting the behavior of the following variables:

- Desired completed visits (DCV), which is the number of visits required in order to avoid a backlog situation (i.e. where the outflow is greater than the inflow).
- Potential completed visits (PCV), which is the number of visits that the system can cater for in a given year. When this number falls below DCV, the backlog accumulates and system pressure builds.
- Effect of schedule pressure on productivity (ESP-P), and the effect of schedule pressure on work year (ESP-WY), both of which indicate the level of response to increasing system pressure.

Figure 4.11 illustrates the difference between the two scenarios. Under *business as usual*, a gap emerges between DCV(1) and PCV(1) as the population ages over time. This clearly shows an unsustainable system, as capacity is always lower than demand. In this simulation, there is no feedback response to increased pressure, so the backlog will also build over time, leading to longer system delays and ultimately poorer health outcomes. What this scenario shows is that the underlying policy for staffing general practitioners is flawed, as taking a crude proportion of the population does not distinguish between the increasing demands as the population ages.

Under the scenario *flexible capacity,* a different result unfolds. Here, the system responds to the increased demands by activating two feedback loops, B_1 and B_2. In this case, both loops respond to the increase in system pressure, and these responses

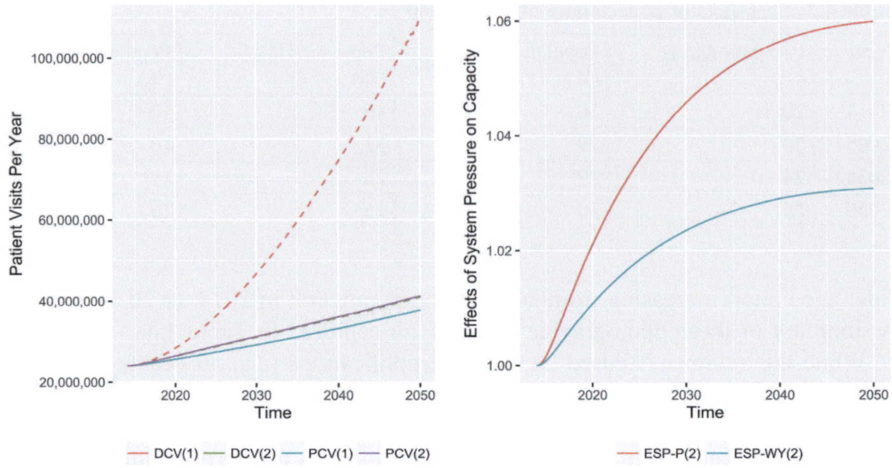

Fig. 4.11 Impact of activating policy options

are also shown in Fig. 4.11. Because of these responses, an increased throughput is realized, and the values for DCV(2) and PCV(2) are in equilibrium. As a result, no backlog builds and the system can absorb the increased demand by (1) extending the work year and (2) enhancing the average productivity of general practitioners.

In summary, this initial model demonstrates a number of important characteristics of system dynamics models. These include:

- Taking a system-wide perspective by modeling different sectors, and identifying causal influences between sectors.
- Modeling at a high level of aggregation in order to engage problem owners. For example, in this problem, a simplified aging chain structure was selected, as well as a basic model of skills generation that did not include a supply line of general practitioners. These are aspects that can be refined in future model iterations.
- Modeling system responses to work pressure through using effects, and maintaining a facility to activate/deactivate these loops in order to support scenario analysis.

Extending the Model

This model version is exploratory and has a number of simplifying assumptions that can be extended across three dimensions: (1) enhancing the endogenous perspective through identifying additional feedback loops, (2) disaggregating stock variables, and (3) accommodating more realistic delays structures. These are now discussed in more detail.

As discussed in Chap. 1, the *endogenous perspective* is fundamental to system dynamics. For a given model, this involves analyzing exogenous variables and considering whether they may be influenced by other system variables, and if such an influence involves the identification of new feedback loops. For example, consider the variables $GPV_{N\text{-}M}$, described in Eqs. (4.37–4.40). These are currently exogenous, as they are constants that do not depend on other stocks or flows. However, the fact that they are exogenous is a modeling decision. The reason was to initially keep the model simple, yet still capture the fact that as the population profile ages, the annual number of GP visits increase. However, these variables could also be viewed as being endogenous. The endogenous-seeking enquiry could be phrased in the following way:

> Is there any other variable in the three-sector model that could influence the annual GP visits?

Those familiar with quality management might identify a candidate influencing variable, which will provide an opportunity to create a link between the delivery and demographic sectors. In response to increasing system pressure, a policy is to increase productivity. This is already captured by (4.57), which shows the systems ability to relieve pressure when demand outpaces supply, by essentially working faster. However, working faster reduces attention to detail, and this can diminish the outcome quality. As a result of this lower quality care, patients may have to revisit their general practitioner as a “rework cycle” is triggered. This can be shown as follows.

↑	Productivity	→	$GPV_{N\text{-}M}$	↑

This adds a new causal link to the model, by making the link between productivity and GP visits. As a consequence, the model now has a new feedback loop, as indicated by following the effect from $GPV_{N\text{–}M}$, on to total GP demand through to system pressure and onto productivity again.

↑	$GPV_{N\text{-}M}$	→	Total GP demand	↑
↑	Total GP demand	→	Patients being treated	↑
↑	Patients being treated	→	Desired completed visits	↑
↑	Desired completed visits	→	System pressure	↑
↑	System pressure	→	Productivity	↑

Interestingly, this is a reinforcing loop as which shows a possible undesirable consequence of reacting to system pressure through increasing productivity (and possibly reducing quality). Further demands are exerted on the system as the rework cycle is activated, and that in turn puts further pressure on practitioners to increase productivity to cope with demands. This is a vicious circle, and its presence is a valuable reminder to policy designers of the limitations to corner-cutting in service-based industries (Oliva 2001).

The second area for model refinement is to *disaggregate stock variables*. This decision is informed by observing the real-world system. For example, a useful question to ask is whether there are differences between sub-cohorts that would alter the model dynamics. This issue will be expanded on in Chap. 5, where a model of disease spread needs to be disaggregated because of the different mixing patterns between age cohorts. In the health care mode, disaggregation can be applied all sectors.

- The model could distinguish between male and female, in order to have more accurate birth rate information based on females within child-bearing age. Demographic models can also be arranged spatially, and this can be important for modeling differences in socio-economic status, and their impact on health outcomes, across a geographic area.
- In the delivery sector, the stock *Patients Being Treated* could be separated into gender and age cohort, as these could have differences in average treatment time required.
- In the supply sector, the general practitioners can be disaggregated in two ways. First, the gender composition of the workforce could be represented, as this is seen as an important factor for GP workforce planning (Lyons and Duggan 2015). Second, career progression, rather than being a single stock, can be reformulated as an aging chain, showing different career stages. This information would be important in order to better identify losses such as the expected retirement rate of GPs.

The third extension that can be applied to the model is to apply more realistic *delay structures* to model the transition of cohorts from one stage to the next. This can apply in two different sectors.

- In the demographic sector, the delay structure could be enhanced by using more intermediary stocks to have a finer granularity of age cohort. For example, rather than having 15 and 25-year delays, a 5-year delay structure could be deployed.
- In the supply sector, supply of general practitioners does not model the decision of higher education institutions to recruit and educate students. This could be achieved with the addition of a supply line stock, and an associated supply line management heuristic. Models from production and distribution systems, such as the beer distribution model (Sterman 1989), could be extended to capture these dynamics.

The modeling process for system dynamics is iterative, where early stage models focus on key stocks, flows, and feedbacks. Further iterations involve elaborating model structures in collaboration with clients. Adopting a sector-based approach is valuable, as this allows the model's complexity to be managed, and also helps to communicate the model to clients.

Summary

This chapter presented a higher order health care model, using key system dynamics structures such as effects, delays and the stock management structure. Further extensions include: identifying exogenous variables that could be endogenous; further elaborating delay structures; and increasing the detail of models through disaggregation. The health theme is continued in Chap. 5 through exploring the spread of infectious diseases, and how the processes of contagion can be successfully modeled in system dynamics.

Exercises

1. For a software organization, the desired number of programmers is one per €100,000 of expected revenue per year. Based on this, construct a stock and flow model of staff recruitment, using the stock management structure, that takes the following into consideration.
   - There are three kinds of programmer: Rookie, Experienced and Expert.
   - All hires are done at the Rookie level, and programmers progress to experienced with an average delay of 50 weeks for rookies, and a delay of 150 weeks before experienced become expert.
   - On average, there is attrition from each programmer category. This is 5 % for Rookies, 2 % for Experienced and 1 % for Expert.
2. Consider the task of software development. Defect density is a measure of the number of defects/line of code (loc) written. Assume that the defect density also depends on the proportion of rookie coders in the organisation. Assuming a reference defect density of 0.05, based on a reference percentage of rookies of 10 %, sketch an overall equation that models the effect of rookie percentage on defect density. Use this equation to build a rework model (stock and flow model with equations) for software construction. Assume that:
   - There is a stock called Code Remaining, which is reduced by Code Completion Rate. This rate reflects the capacity of a software team, where the team is made up of Rookies and Experienced coders.
   - Rookies become experienced after a first order time delay of 50. Rookie productivity is 30 loc/coder/day, whereas experienced productivity is 150 loc/coder/day.
   - The stock of Completed Code can then flow into Fully Working Code, although a percentage flows into Undiscovered Code Errors. After a first order time delay, these errors flow back into the stock of Code Remaining, and this completes the rework cycle.

References

Cachon G, Terwiesch C (2009) Matching supply with demand, vol 2. McGraw-Hill, Singapore

Coyle RG (1996) System Dynamics Modelling: a Practical Approach. CRC Press

Forrester JW (1961) Industrial Dynamics. MIT Press, Cambridge MA. (Reprinted by Pegasus Communications: Waltham, MA)

Forrester JW (1968) Market growth as influenced by capital investment. Ind Manag Rev 9(2):83

Forrester JW (1987) Lessons from system dynamics modeling. Syst Dyn Rev 3(2):136–149

Hirsch G, Homer J, Tomoaia-Cotisel A (2013) System dynamics applications to health and health care. Syst Dyn Rev, Special Virtual Issue. http://onlinelibrary.wiley.com/journal/10.1002/(ISSN)1099-1727/homepage/VirtualIssuesPage.html#Health. Accessed 20 July 2015

Lyons GJ, Duggan J (2015) System dynamics modelling to support policy analysis for sustainable health care. J Simul 9(2):129–139

Oliva R (1996) A dynamic theory of service delivery: implications for managing service quality. Doctoral dissertation, Massachusetts Institute of Technology

Oliva R, Sterman JD (2001) Cutting corners and working overtime: quality erosion in the service industry. Manage Sci 47(7):894–914

Sterman JD (1989) Modeling managerial behavior: misperceptions of feedback in a dynamic decision making experiment. Manage Sci 35(3):321–339

Sterman JD (2000) Business dynamics: systems thinking and modeling for a complex world. Irwin/McGraw-Hill, Boston

Chapter 5
Diffusion Models

The historical and epidemiological literature abound with accounts of infectious diseases invading human communities and of the concomitant effects on population abundance, social organization, and the unfolding patterns of historical events.
R. Anderson, R. May. *Infectious Diseases of Humans: Dynamics and Control (1992).*

Abstract This chapter focuses on diffusion, which is a common feature of many social and biological systems. Innovative consumer products frequently "take off" and "go viral", with sales driven by the word of mouth effect, as their adoption spreads through a population. Infectious diseases can transmit rapidly through a population, accelerating from seemingly low incidence levels, to sizable numbers in a short space of time. Here, the focus is on models of infectious diseases. These have an important decision support function for public health professionals faced with challenge of responding to an infectious disease outbreak. The first model is the classic SIR structure, which divides the population into those who are susceptible, infected and recovered. This model is then extended to cater for multiple age cohorts, so that diverse mixing patterns can be simulated. Finally, a scalable R model of infectious diseases is introduced, combining matrix operations with vectorized differential equations.

Keywords SIR model · Policy analysis · Disaggregate SIR model · Model vectorization

The SIR Model

System dynamics research on modeling infectious disease transmission includes an analysis of the consequences of antiretroviral therapy (Dangerfield et al. 2001), and policies for the global management of poliomyelitis (Thompson and Tabbens 2008). Because of the potential negative impact of these processes, decision makers need to model diffusion and its properties. Stock and flow models can support this

J. Duggan, *System Dynamics Modeling with R*, Lecture Notes in Social Networks,
DOI 10.1007/978-3-319-34043-2_5

objective. A widely-used model of infectious disease transmission is the SIR structure, and this divides a population into three stocks:

- *Susceptible* (*S*), models those in the population who have not acquired a contagious disease, but who are at risk of being infected, should they encounter an infected person.
- *Infected* (*I*), contains people who have a disease and can spread this to others through contact.
- *Recovered* (*R*), captures those who are immune to the disease, and will not transmit it through daily contacts with other members of the population.

A feature of this model is that individuals reside in any one of the three stocks. For example, an entire population could be susceptible to a new strain of influenza. If an individual becomes infected, this can lead to a "chain reaction" of infection, as a person infects others, who in turn move from the susceptible stock to the infected stock. Infected people recover after a time delay, and move from being infected to being recovered. These transitions between stocks are governed by the flows in the SIR model.

The overall stock and flow model is illustrated in Fig. 5.1, where the three stocks are susceptible (5.1), infected (5.2) and recovered (5.3). The flows determine the rate at which individuals move between the stocks, and one of these, the recovery rate (RR) is a first order delay, introduced earlier in Chap. 4.

$$Susceptible\,(S) = INTEGRAL(-IR, 99999) \tag{5.1}$$

$$Infected\,(I) = INTEGRAL(IR - RR, 1) \tag{5.2}$$

$$\text{Recovered}(R) = INTEGRAL(RR, 0) \tag{5.3}$$

To model disease transmission through the infection rate (IR), an important definition is known as the *force of infection*, denoted as lambda (λ). This is defined as the rate at which susceptible individuals become infected per unit time

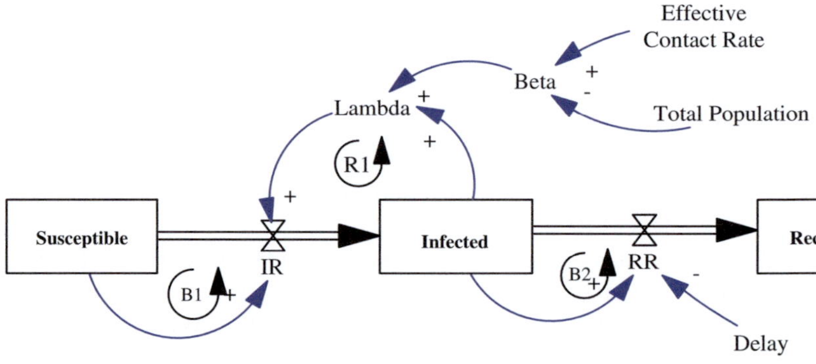

Fig. 5.1 The SIR model of contagion

(Vynnycky and White 2010). The force of infection is proportional to the number of infected people. This is intuitive, as the greater the number of infected people, the greater the likelihood that more susceptible people will become infected.

This feedback dynamic can be confirmed by calculating the loop polarity in the SIR model. As the number of infected cases increase, so too does lambda. An increase in lambda leads to an increased in the infection rate (IR), which in turn leads to higher numbers of infected. This is a reinforcing process, and the positive feedback loop can quickly dominate the model behavior and so drive the exponential growth processes associated with the outbreak of a contagious disease.

↑	Infected	→	Lambda	↑
↑	Lambda	→	IR	↑
↑	IR	→	Infected	↑

The force of infection is described in Eq. (5.4), where beta (β) is a constant that is used to quantify the strength of disease transmission, via contacts.

$$Lambda(\lambda) = Beta(\beta) \times I \tag{5.4}$$

Beta is formally defined as the per capita rate at which two specific individuals come into effective contact per unit time (Vynnycky and White 2010), and is defined in Eq. (5.5), where the total population is the sum of all stocks (5.10).

$$Beta(\beta) = \frac{C_E}{N} \tag{5.5}$$

The value of β is based on the effective contact rate (C_E) between members of the population. An effective contact is a contact that is *sufficient to lead to transmission if it occurs between an infectious and susceptible person* (Vynnycky and White 2010). C_E (5.6) is based on an estimate of contact frequency within the population, and the likelihood that such an interaction leads to infection. For example, if the average contact rate in the population is 8 people/person/day and the chances of an infectious person infecting a susceptible person is ¼, then the effective contact rate is the product of these terms (i.e. 2 people/person/day). For this initial model, this value of 2 is used.

$$Effective\ Contact\ Rate\ (CE) = 2 \tag{5.6}$$

Beta is also known as the *transmission parameter* and can depend on a number of factors, including age and geographic setting. For example, because of different contact rates, beta values are likely to be higher in children than in adults, and also for individuals living in urban settings as opposed to rural areas.

Mitigation strategies can also reduce β, for instance, closing down schools during an influenza outbreak reduces contacts between children, and hence can slow down the spread of infection. Once the values for β and λ are known, the infection rate (IR) can be calculated, and this is the product of the force of infection and the number of susceptible individuals in the population, as shown in Eq. (5.7).

$$IR = S \times \lambda = \beta SI \tag{5.7}$$

There are a number of interesting observations that can be made about Eq. (5.7), particularly when considering the conditions under which disease transmission will not occur.

- If there are no susceptible people, there can be no infections, as $S = 0$.
- If there are no infected people circulating in the population, there can be no infections, as $I = 0$.
- If there are no effective contacts in the population (i.e. $C_E = 0$), then $\beta = 0$, and there can be no new infections.

While these may seem self-evident, the points demonstrate the underlying robustness of the SIR model in that its transmission equation (i.e. the value IR) maps well onto the conditions necessary for disease spread. The SIR model includes a second flow equation for the recovery rate (RR), which governs the outflow from the infected stock. With this Eq. (5.8), individuals are being continually removed from the infected stock by means of a first order delay, with time constant D (5.9).

$$RR = \frac{I}{D} \tag{5.8}$$

$$Delay\,(D) = 2 \tag{5.9}$$

$$Total\ Population\,(N) = Susceptible + Infected + Recovered \tag{5.10}$$

Before exploring the SIR model implementation in R, and the simulation output, it is useful to see how these equations operate. Consider the following scenario.

> In a town of size 100,000, one person has been infected with a new strain of influenza. Therefore $S = 99999$, $I = 1$ and $R = 0$. Assume a recovery delay of 2 days, and an effective contact rate $C_E = 2$. Calculate the values for β and λ, and the initial number of susceptible people that become infected.

To solve this, the following equations are used.

- $\beta = \frac{C_E}{N} = \frac{2}{100{,}000} = 2 \times 10^{-5}$
- $\lambda = \beta I = 2 \times 10^{-5} \times 1 = 2 \times 10^{-5}$
- $IR = \lambda S = 2 \times 10^{-5} \times 99{,}999 = 1.99998$

Therefore, with one person infected the model will generate a further 1.99998 people to leave the susceptible stock, and join the infected stock. This sequence of calculations is performed through each iteration of the simulation, in order to determine how many susceptible people become infected. The R code for the SIR model is now presented.

First, the overall simulation parameters are coded, along with the three stocks, and the auxiliaries. In this case, the population is set as a constant, although it could also be defined as a variable inside the model function.

```
START<-0; FINISH<-20; STEP<-0.125

simtime <- seq(START, FINISH, by=STEP)
stocks  <- c(sSusceptible=99999,sInfected=1,sRecovered=0)
auxs <- c(aTotalPopulation=100000, aEffective.Contact.Rate=2,
           aDelay=2)
```

The model function solves the equations in the correct sequence, starting with β (aBeta) and λ (aLambda), progressing through the two flows (fIR and fRR), and finishing by evaluating the three integrals, and returning the list of variables to the deSolve **ode()** routine.

```
model <- function(time, stocks, auxs){
  with(as.list(c(stocks, auxs)),{
    aBeta    <- aEffective.Contact.Rate / aTotalPopulation
    aLambda <- aBeta * sInfected

    fIR      <- sSusceptible * aLambda
    fRR      <- sInfected / aDelay

    dS_dt    <- -fIR
    dI_dt    <- fIR - fRR
    dR_dt    <- fRR

    return (list(c(dS_dt, dI_dt, dR_dt),
                  IR=fIR, RR=fRR,Beta=aBeta,Lambda=aLambda,
                  CE=aEffective.Contact.Rate))
  })
}
```

The function is then called, and the output analyzed.

```
o<-data.frame(ode(y=stocks, times=simtime, func = model,
                   parms=auxs, method="euler"))
```

The simulation output is displayed in Fig. 5.2. This shows how the infection rate equation replicates a classic infection outbreak, with a low initial value which accelerates as more people get infected. The infection rate curve peaks before

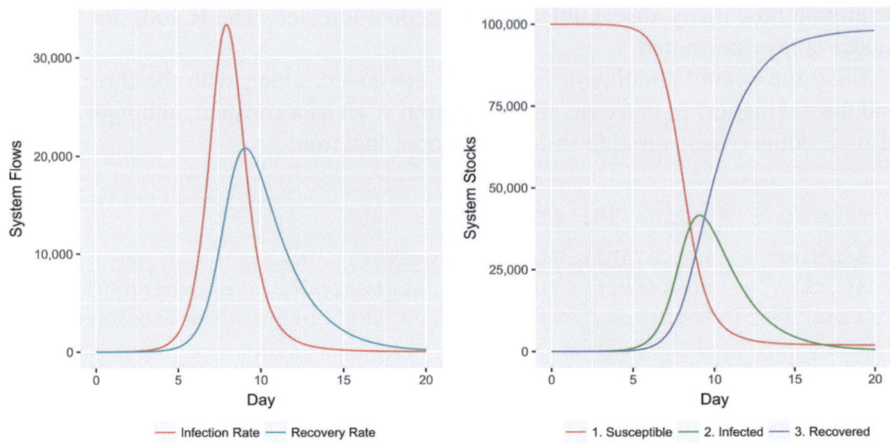

Fig. 5.2 Model output showing contagion effect

declining, as then there are fewer people available to infect. While the beta value stays constant for the entire simulation, the force of infection is continually changing, as more people accumulate in the infected stock. The stocks also give a clear indication of the disease dynamics, as the number susceptible is initially very high, but as the number of infected rise and gather momentum, the stock of susceptible falls rapidly (as the positive feedback loop dominates).

The infected stock, which models disease prevalence, is also of particular importance to health officials. The simulated peak values can give a good indication of demand surges on public health services (e.g. visits to general practitioners, hospital admissions, and demands for intensive care unit facilities). This can support emergency response planning in scenarios such as low supply of vaccines, which is a likely scenario with the outbreak of, for example, a new strain of a highly contagious influenza virus.

However, while the model is useful for exploring disease dynamics, and demonstrates the power of a positive feedback loop to quickly spread contagion, it does not yet provide the facility for policy analysis. In real-world epidemic scenarios public health officials take action to reduce the impact of disease spread, and these interventions can include:

- Vaccination, where susceptible people are administered a vaccine to ensure immunity.
- Quarantine, where individuals who are infected remove themselves from contact with others, in order to reduce the transmission rate.
- Social distancing, where contact rates are reduced through actions such as school closures, and the cancellation of public gatherings.

Two of these policy responses are now considered by extending the initial SIR model, and adding three new flows, with one additional stock.

Policy Exploration with the SIR Model

The SIR model simulates disease transmission outcomes if no decisions were taken to counter the impact of disease spread. However, an important idea in system dynamics is that policy responses are modeled, in order to evaluate the impact of different interventions. In many cases policy interventions involve focusing on flows, and considering what can be done to either reduce or increase a flow. For the SIR model, the key flow to constrain is the infection rate (5.7), and identify what interventions can reduce this value.

Figure 5.3 shows two possible interventions that can reduce the infection rate. There are to:

- Introduce vaccinations, so that people can flow directly from susceptible to recovered, without infecting other individuals. From an equation perspective, introducing a vaccination flow will reduce the susceptible cohort, and this is turn will reduce the infection rate (5.7).
- Introduce quarantine, where a proportion of those infected self-isolate and move to a new stock (quarantine). This reduces the number of infectious people in circulation (i.e. the infected stock), which in turn reduces the force of infection λ (5.4), and subsequently the infection rate.

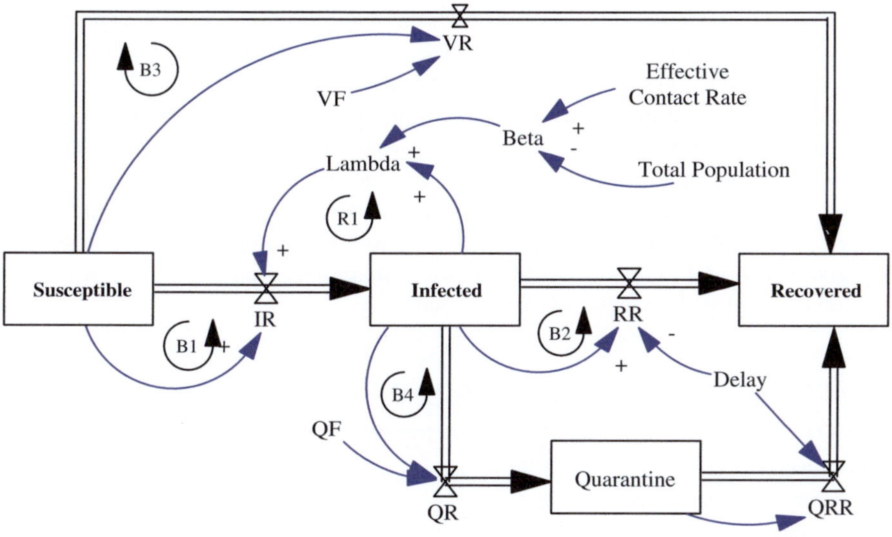

Fig. 5.3 The SIR model with containment policy options

To model these two policy options the model equations are updated to include new flows, auxiliaries and a stock. Outflows are added to the susceptible stock (5.11) in the form of a vaccination rate (VR), and to the infected stock (5.12) through a quarantine rate (QR). An outflow is required for the quarantine stock (5.13), and this serves as an inflow to the recovered stock (5.14), along with the flow VR.

$$Susceptible\,(S) = INTEGRAL(-(IR + VR), 99999) \tag{5.11}$$

$$Infected\,(I) = INTEGRAL(IR-(RR + QR), 1) \tag{5.12}$$

$$Quarantine\,(Q) = INTEGRAL(QR - QRR, 0) \tag{5.13}$$

$$Recovered\,(R) = INTEGRAL(QRR + RR + VR, 0) \tag{5.14}$$

Three new flows are specified as first-order delay processes. The vaccination rate (5.15) is a fixed proportion of the susceptible stock, and the quarantine rate is a fraction of the infected population (5.16). The quarantine recovery rate (5.17) is a first order delay process based on the disease duration, similar to Eq. (5.8).

$$Vaccination\ Rate\,(VR) = S \times Vaccination\ Fraction \tag{5.15}$$

$$Quarantine\ Rate\,(QR) = I \times Quarantine\ Fraction \tag{5.16}$$

$$Quarantine\ Recovery\ Rate\,(QRR) = \frac{I}{D} \tag{5.17}$$

With the model reformulated, scenario analysis can be performed. For this, it is useful to focus on a specific variable (often called the *variable of interest*), and compare the behavior of this variable under a range of different policy responses. In this case, the variable of interest is *infected*, as this is what public health officials want to minimize. Four scenarios are summarized in Table 5.1, which include permutations of combining vaccination and quarantine. When generating scenarios it is important to provide a base case scenario where no intervention is taken, as this can then be benchmarked with the results of other scenarios.

The choice of fractional values is based on two assumptions: (1) there is a limited supply of vaccines so that only 5 % of the population can be vaccinated on any given day, and (2) it is assumed that the quarantine fraction is low, with only 5 % of infected people self-isolating on each day.

Table 5.1 Scenarios exploring mitigation policies

Scenarios	Vaccination fraction (VF)	Quarantine fraction (QF)
(1) No interventions	0.00	0.00
(2) Vaccinate, no quarantine	0.05	0.00
(3) Quarantine, no vaccination	0.00	0.05
(4) Vaccinate and quarantine	0.05	0.05

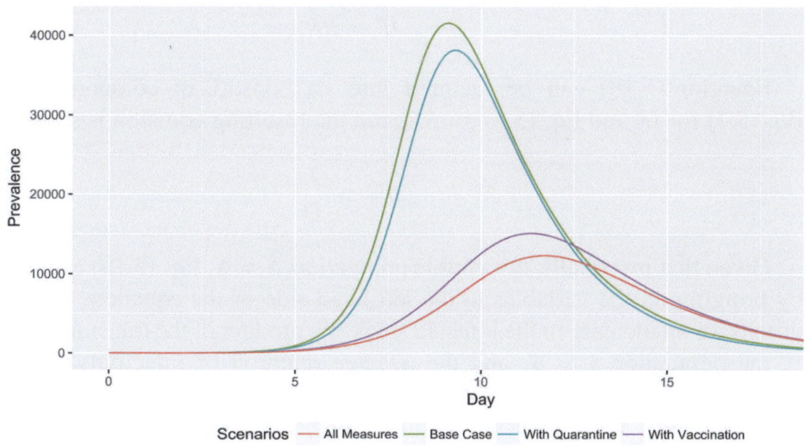

Fig. 5.4 Comparison of policy responses to outbreak

The simulation output is captured in Fig. 5.4, and shows the practical benefits of using simulation to explore a range of responses.

- For the base case, with no policies enacted, the peak is highest, and also occurs at the earliest time in the simulation. This models the worst case scenario, where infection rates increase rapidly, and would lead to a considerable strain on a public health system.
- The quarantine policy, where only 5 % of infected people are isolated, does not have a significant impact on the prevalence peak. This is because that the rate of removal from the infected stock is not sufficient to stop the disease spread, as there are still sufficient quantities of infected people in circulation to ensure that the virus spreads widely.
- Vaccination results in a significant impact on the prevalence, as the peak of the curve is smaller, and the time of the peak is pushed out, thereby reducing the impact on health services.
- The combination of vaccination and quarantine lead to the most desirable result, as the peak is reduced, and the peak time pushed further into the future.

A deeper understanding of disease dynamics can be obtained by performing a mathematical analysis on elements of the original SIR model. This is achieved by focusing on the inflow and outflow of the infected stock. A basic principle of stock and flow system is that for a stock to rise, the inflow must exceed the outflow. For the SIR model, this means that the prevalence will rise if the infection rate is higher than the recovery rate, and this is shown in Eq. (5.18).

$$IR > RR \tag{5.18}$$

Equation (5.18) can be factored into its constituent components, based on Eq. (5.7) for *IR* and Eq. (5.8) for *RR*, and the resulting equation is shown in (5.19).

$$\beta SI > \frac{I}{D} \tag{5.19}$$

Given that in a totally susceptible population, $S = N$, Eq. (5.19) can be simplified by bringing all the variables to the left hand side of the equation, so that the conditions for an increase in the infected stock is a product of the transmission parameter β, the population size N, and the average duration of infectivity D (5.20). This equation can also be represented in terms of effective contacts (5.21) by replacing the value of β with its equivalent form described earlier in the chapter (5.5).

$$\beta ND > 1 \tag{5.20}$$

$$C_E D > 1 \tag{5.21}$$

From a policy analysis perspective, these equations are important as they represent the conditions under which an epidemic will occur in a population. If the overall condition is true, then the infection rate (inflow) will exceed the recovery rate (outflow), and the number of infected will rise. For example, returning to the previous example of a town of 100,000 inhabitants, with $D = 2$, and $C_E = 2$, we can see that Eq. (5.21) will evaluate as $2 \times 2 = 4$. As this value is greater than 1, an epidemic will occur, and this is confirmed by the simulation output shown earlier in Fig. 5.2.

In the context of infectious disease dynamics and control, an additional variable is widely used amongst epidemiologists. This is known as the basic reproduction number R_0, which is the average number of secondary infectious persons resulting from one infectious person being introduced to a totally susceptible population (Anderson and May 1992). In the SIR model, this can be formulated as the product of the effective contact rate and the average duration of infectiousness, as shown in Eq. (5.22).

This equation is intuitive:

- if an infectious individual has 2 effective contacts per day, and
- they are infectious for 2 days,
- they will generate $2 \times 2 = 4$ secondary infections, and,
- therefore so the value of R_0 is 4.

$$R_0 = C_E D = \beta ND \tag{5.22}$$

In epidemiology practice, R_0 can be calculated based on individual level contact tracing, estimations from the early growth curves of an infection, and through the

Table 5.2 Typical R_0 values for certain infectious diseases

Infection	Influenza	Measles	Mumps	Pertussis
R_0	2–4	12–18	4–7	12–17

use of model calibration (Breban et al. 2007). The value of R_0 varies according to the infection, and typical values for a range of infectious diseases are shown in Table 5.2.

Finally, if the values for the reproduction number (R_0), and the infectious period (D) are both known, then the transmission parameter (β) can be directly calculated, as shown in Eq. (5.23).

$$\beta = \frac{R_0}{ND} \tag{5.23}$$

A Disaggregate SIR Model

A feature of the initial SIR model is that the population mixes randomly, and the chance that any two individuals meet is equally likely. Because of this, no distinction is made between any age cohort, which is unrealistic in certain scenarios. For infectious disease spread, the case for a disaggregate system dynamics model is compelling, given the evidence of age-dependent mixing. A number of empirical studies have confirmed non-random mixing in populations, including data on the transmission of tuberculosis (Borgdorff et al. 1999).

Furthermore, a significant study of contact patterns across Europe showed that mixing patterns are highly assortative (i.e. "with-like") with age (Mossong et al. 2008). With this in mind, the original SIR model can be disaggregated into three distinct age cohorts: young (Y), adult (A) and elderly (E).

The disaggregate SIR model is shown in Fig. 5.5, and has a number of features:

- It replicates the SIR stock and flow structure for each age cohort.
- Each age cohort has its own force of infection, which is determined by interactions with all model cohorts.
- The transmission parameter β is disaggregated to model pair-wise interactions amongst all the cohorts. This is intuitive, as parameters are identified to capture the full set of interactions between cohort members, and therefore the "with-like" reality of social mixing can be modeled.

For this model, the overall population of 100,000 is split into an arbitrary ratio of 1:2:1 between the three cohorts $\{Y, A, E\}$, and the susceptible equations (S_Y, S_A, and S_E) are shown in (5.24–5.26), along with their initial values.

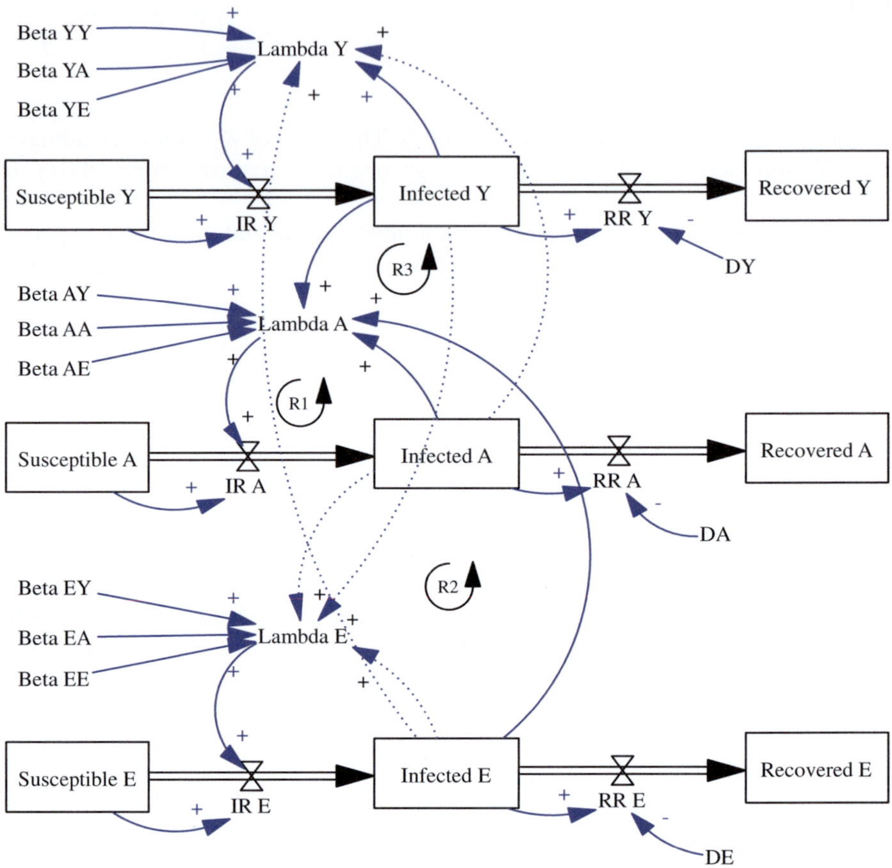

Fig. 5.5 A disaggregated SIR model (three cohorts)

$$Susceptible\ Y\ (S_Y) = INTEGRAL(-IR_Y, 24999) \tag{5.24}$$

$$Susceptible\ A\ (S_A) = INTEGRAL(-IR_A, 50000) \tag{5.25}$$

$$Susceptible\ E\ (S_E) = INTEGRAL(-IR_E, 25000) \tag{5.26}$$

The subsequent stock equations for the infected cohorts (I_Y, I_A, and I_E) are specified in Eqs. (5.27–5.29), and in this case only one person, from the young cohort, is initially infected.

$$Infected\ Y\ (I_Y) = INTEGRAL(IR_Y - RR_Y, 1) \tag{5.27}$$

$$Infected\ A\ (I_A) = INTEGRAL(IR_A - RR_A, 0) \tag{5.28}$$

$$Infected\ E\ (I_E) = INTEGRAL(IR_E - RR_E, 0) \tag{5.29}$$

The final set of stock equations model the recovered cohorts (R_Y, R_A, and R_E), and given that the simulation is exploring the impact of a new virus on a totally susceptible population, the initial value of all these stocks, listed in Eqs. (5.30–5.32), is zero.

$$Recovered\ Y\ (R_Y) = INTEGRAL(RR_Y, 0) \tag{5.30}$$

$$Recovered\ A\ (R_A) = INTEGRAL(RR_A, 0) \tag{5.31}$$

$$Recovered\ E\ (R_E) = INTEGRAL(RR_E, 0) \tag{5.32}$$

While the stock equations are relatively straightforward, the structure of the force of infection equations is more challenging. The general form of the force of infection for a cohort i in a population of N cohorts is shown in Eq. (5.33).

$$\lambda_i = \sum_{j=1}^{N} \beta_{ij} I_j \tag{5.33}$$

The force of infection for a cohort is influenced by interactions with all other cohorts. The notation for β_{ij} is significant. This can be interpreted as the transmission parameter *from* an infectious cohort j *to* a susceptible cohort i.

The force of infections are now formulated. For the first cohort, the force of infection for the young cohort λ_Y (Eq. 5.34) is the weighted sum of the force of infections from each cohort interaction. The terms β_{YY}, β_{YA}, and β_{YE} model the transmission parameters for each cohort interaction, and these are multiplied by the relevant number of infected people in each cohort.

$$Lambda\ Y\ (\lambda_Y) = \beta_{YY} I_Y + \beta_{YA} I_A + \beta_{YE} I_E \tag{5.34}$$

The values for these pair-wise transmission parameters are defined in Eqs. (5.35–5.37), and are expressed in terms of the per-capita effective contacts, where the total cohort population for the *susceptible cohort* is used in the denominator term. An estimate of the effective contacts between young and young, young and adult and young and elderly is specified in Eqs. (5.38–5.40). Notice that the highest value is for "with-like" mixing, where CE_{YY} has a value of 3.0. This models a scenario where higher effective contacts occur between young people, based on the observation that teenagers may well form the transmission backbone of future epidemics (Glass et al. 2008).

$$\beta_{YY} = CE_{YY}/N_Y \tag{5.35}$$

$$\beta_{YA} = CE_{YA}/N_Y \tag{5.36}$$

$$\beta_{YE} = CE_{YE}/N_Y \tag{5.37}$$

$$CE_{YY} = 3.0 \tag{5.38}$$

$$CE_{YA} = 2.0 \tag{5.39}$$

$$CE_{YE} = 1.0 \tag{5.40}$$

In a similar manner, the force of infection for the remaining cohorts, λ_A and λ_Y are defined in Eqs. (5.41 and 5.42). Again, each of these equations illustrate that a cohort's force of infection is influenced by all other infected cohorts in the model.

$$Lambda\ A\ (\lambda_A) = \beta_{AY}I_Y + \beta_{AA}I_A + \beta_{AE}I_E \tag{5.41}$$

$$Lambda\ E\ (\lambda_E) = \beta_{EY}I_Y + \beta_{EA}I_A + \beta_{EE}I_E \tag{5.42}$$

The remaining transmission parameters (Eqs. 5.43–5.48) and effective contacts (5.49–5.54) are also defined, and so the specification of all equations for the force of infection variables are completed.

$$\beta_{AY} = CE_{AY}/N_A \tag{5.43}$$

$$\beta_{AA} = CE_{AA}/N_A \tag{5.44}$$

$$\beta_{AE} = CE_{AE}/N_A \tag{5.45}$$

$$\beta_{EY} = CE_{EY}/N_E \tag{5.46}$$

$$\beta_{EA} = CE_{EA}/N_E \tag{5.47}$$

$$\beta_{EE} = CE_{EE}/N_E \tag{5.48}$$

$$CE_{AY} = 2.0 \tag{5.49}$$

$$CE_{AA} = 2.0 \tag{5.50}$$

$$CE_{AE} = 1.0 \tag{5.51}$$

$$CE_{EY} = 1.0 \tag{5.52}$$

$$CE_{EA} = 1.0 \tag{5.53}$$

$$CE_{EE} = 0.5 \tag{5.54}$$

For clarity, it is recommended that the effective contact values are displayed in matrix format, so that the effective contact interactions can be communicated in a user-friendly manner. This is shown in Table 5.3, and the matrix values are

Table 5.3 Inter-cohort effective contact matrix

		From		
		Young	Adult	Elderly
To	Young	3.0	2.0	1.0
	Adult	2.0	2.0	1.0
	Elderly	1.0	1.0	0.5

symmetrical, as the effective contacts from cohort A to B, are the same as the effective contact from B to A.

Based on these equations for the forces of infection, the infection rates for each cohort are specified. These are the product of the force of infection times the cohort susceptible stock, as shown in Eqs. (5.55–5.57).

$$IR_Y = \lambda_Y S_Y \tag{5.55}$$

$$IR_A = \lambda_A S_A \tag{5.56}$$

$$IR_E = \lambda_E S_E \tag{5.57}$$

Finally, the flow equations for each cohort's recovery rate are defined, and these are first order delay structures, which the outflow is proportional to the value in the infected stock. These equations are documented in (5.58–5.60), and the time constants shown in Eqs. (6.61–6.63). While the time constants are the same for this model, it is useful to have three separate variables, as it allows the modeler to experiment with different delay values across the three cohorts.

$$RR_Y = \frac{I_Y}{D_Y} \tag{5.58}$$

$$RR_A = \frac{I_A}{D_A} \tag{5.59}$$

$$RR_E = \frac{I_E}{D_E} \tag{5.60}$$

$$D_Y = 2.0 \tag{5.61}$$

$$D_A = 2.0 \tag{5.62}$$

$$D_E = 2.0 \tag{5.63}$$

This disaggregate SIR model is now implemented in R, in what turns out to be a surprisingly straightforward, and scalable, implementation, given the opportunity to use matrix algebra and vectorized equations as part of the solution.

A Vectorized Disaggregated SIR Model in R

An interesting property of the SIR cohort model is that the force of infection equations can be presented in a generalized form, and matrix algebra can be used for specifying equations in the system dynamics model. Specifically, the general equation for the force of infection, described in (5.33), can be re-formulated as a matrix multiplication operation (5.64).

$$\begin{bmatrix} \lambda_1 \\ \vdots \\ \lambda_N \end{bmatrix} = \begin{bmatrix} ce_{11}/N_1 & \cdots & ce_{1n}/N_1 \\ \vdots & \ddots & \vdots \\ ce_{n1}/N_N & \cdots & ce_{nn}/N_N \end{bmatrix} \begin{bmatrix} I_1 \\ \vdots \\ I_N \end{bmatrix} \tag{5.64}$$

where:

- There are N cohorts to be modeled. For SIR purposes, the cohorts are usually disaggregated by age, but they could also be divided by geographic area.
- The force of infection for each cohort i is given by the value λ_i
- The effective contact rates are modeled between each cohort ce_{ij}, where i is the susceptible cohort and j is the infectious cohort.
- The cohort sub-population for cohort i values are denoted by N_i
- The number of infected for each cohort i is given by the value I_i

Given this general equation, the force of infections (5.34, 5.41 and 5.42) from the earlier model can be represented in matrix form (5.65).

$$\begin{pmatrix} \lambda_Y \\ \lambda_A \\ \lambda_E \end{pmatrix} = \begin{pmatrix} CE_{YY}/N_Y & CE_{YA}/N_Y & CE_{YE}/N_Y \\ CE_{AY}/N_A & CE_{AA}/N_A & CE_{AE}/N_A \\ CE_{EY}/N_E & CE_{EA}/N_E & CE_{EE}/N_E \end{pmatrix} \begin{pmatrix} I_Y \\ I_A \\ I_E \end{pmatrix} \tag{5.65}$$

Therefore, at the simulation outset, with $I_Y = 1$, the initial force of infection values are calculated and shown in Eq. (5.66).

$$\begin{pmatrix} \lambda_Y \\ \lambda_A \\ \lambda_E \end{pmatrix} = \begin{pmatrix} 3.0/25,000 & 2.0/25,000 & 1.0/25,000 \\ 2.0/50,000 & 2.0/50,000 & 1.0/50,000 \\ 1.0/25,000 & 1.0/25,000 & 0.5/25,000 \end{pmatrix} \begin{pmatrix} 1 \\ 0 \\ 0 \end{pmatrix} = \begin{pmatrix} 0.000120 \\ 0.000040 \\ 0.000040 \end{pmatrix} \tag{5.66}$$

R has a number of interesting features that can be used to take full advantage of these matrix equations.

- R supports the full set of matrix representation and operations so that Eq. 5.66 can be solved.
- The deSolve function supports vectorized operations, so that large sets of equations can be solved using vectors.

The R code for a disaggregate SIR model is now presented, based on the three cohort example. It's important to highlight that this code can cater for a much higher number of cohorts. This could be particularly useful if a model builder was deploying these equations to model disease spread over a wide geographic area, or across a wider range of cohorts.

Initially, the model constants are listed. These include two new constants, *NUM_COHORTS*, which captures the number of age cohorts in the model, and *NUM_STATES* which specify the number of stocks in the main disease transmission model. In this case the number of cohorts (Y, A and E) has the same value as the number of disease states (S, I and R). The simulation time vector is also defined, and runs from day 0 to day 20, with a time step of 0.125.

```
START<-0; FINISH<-20; STEP<-0.125;
NUM_COHORTS<-3; NUM_STATES<-3

simtime <- seq(START, FINISH, by=STEP)
```

Next, the effective contact values are recorded in a matrix structure, using R's **matrix()** function. These values could also be read from a spreadsheet or database. For this example, the values summarized earlier in Table 5.3 are used.

```
CE  <- matrix(c(3.0, 2.0, 1.0,
                2.0, 2.0, 1.0,
                1.0, 1.0, 0.5), nrow=3,ncol=3,byrow=TRUE)

> CE
     [,1] [,2] [,3]
[1,]    3    2  1.0
[2,]    2    2  1.0
[3,]    1    1  0.5
```

The total number of individuals in each cohort are also specified, in standard vector format.

```
CohortPopulatons<- c(Young=25000,Adult=50000,Elderly=25000)

> CohortPopulatons
[1] 25000 50000 25000
```

Based on these two variables, a matrix of beta values can be calculated, using the equations specified in (5.65), by simply dividing the contact matrix by the population vector. The values for the beta matrix can also be viewed through the R console, and this matrix will be used to calculate the forces of infection.

```
beta <- CE/CohortPopulatons

> beta
        [,1]  [,2]  [,3]
[1,] 0.00012 8e-05 4e-05
[2,] 0.00004 4e-05 2e-05
[3,] 0.00004 4e-05 2e-05
```

Before the solver function is called, the full set of model integrals need to be specified. In this case, there are nine stock variables, and they are listed in a single vector. The sequence is important, and the stocks are grouped by type, not by cohort, and their initial values specified. The reason for this ordering will become clear when the overall solver equations are presented.

```
stocks <- c(SusceptibleY=24999, SusceptibleA=50000,
            SusceptibleE=25000, InfectedY=1,
            InfectedA=0, InfectedE=0,
            RecoveredY=0, RecoveredA=0,
            RecoveredE=0)
```

Next, the delays are assigned in a vector. In this model, there is no requirement for further auxiliaries to the model, so that value is set to NULL.

```
delays <- c(DY=2.0, DA=2.0, DE=2.0)

auxs <-NULL
```

The model function is now specified, and it contains a number of interesting features that support extension of the SIR disaggregate model. In particular, this function is scalable, and can work for any number of cohorts, once there are sufficient computing resources available.

```
model <- function(time, stocks, auxs){
  with(as.list(c(stocks, auxs)),{

    states<-matrix(stocks, nrow=NUM_COHORTS, ncol=NUM_STATES)

    Susceptible <- states[,1]
    Infected    <- states[,2]
    Recovered   <- states[,3]

    Lambda      <- beta %*% Infected

    IR          <- Lambda * Susceptible
    RR          <- Infected / delays

    dS_dt       <- -IR
    dI_dt       <- IR - RR
    dR_dt       <- RR

    return (list(c(dS_dt, dI_dt, dR_dt)))
  })
}
```

The first step is to convert the vector of incoming stock values into a matrix, where each column in the matrix contains those values for a common stock (e.g. S_Y, S_A and S_E). The matrix function in R can transform a vector into a two dimensional matrix as follows.

```
states<-matrix(stocks, nrow=NUM_COHORTS, ncol=NUM_STATES)
```

For example, the first time the model is run, the values for this matrix are shown below.

```
> states
      [,1] [,2] [,3]
[1,] 24999    1    0
[2,] 50000    0    0
[3,] 25000    0    0
```

This shows that the matrix is simply a different way to represent all the stock variables, and has been filled in column order. This is useful, because each column now represents a model state for each cohort. The first column represents S_Y, S_A, and S_E, the second column represents I_Y, I_A, and I_E, and the third column represents R_Y, R_A, and R_E. Three one-column matrices can now be extracted from each of the matrix columns to obtain all of the state values for each model stocks. These are conveniently organized by stock type, and will be used in later calculations.

```
Susceptible <- states[,1]
> Susceptible
[1] 24999 50000 25000
Infected    <- states[,2]
> Infected
[1] 1 0 0
Recovered   <- states[,3]
> Recovered
[1] 0 0 0
```

With all the state information available, and the beta values already calculated, R's matrix library is used to calculate all the force of infection values. The matrix multiplication operator %*% is used to implement Eq. (5.65), and produce a one-column matrix of lambda values, which is the same result that was calculated earlier in Eq. (5.66). The rows in this one-column matrix represent λ_Y, λ_A and λ_E.

```
Lambda      <- beta %*% Infected
> Lambda
        [,1]
[1,] 0.00012
[2,] 0.00004
[3,] 0.00004
```

All necessary information is now available to calculate the flows, using element-wise multiplication for IR, and element-wise division for RR. The vector IR represents IR_Y, IR_A, and IR_E, and the vector RR accounts for RR_Y, RR_A, and RR_E.

```
IR         <- Lambda * Susceptible
> as.vector(IR)
[1] 2.99988 2.00000 1.00000
RR         <- Infected / delays
> as.vector(RR)
[1] 0.5 0.0 0.0
```

With the flows available, the integrals are then evaluated, and are then returned to the solver.

```
dS_dt  <- -IR
> as.vector(dS_dt)
[1] -2.99988 -2.00000 -1.00000
dI_dt  <- IR - RR
> as.vector(dI_dt)
[1] 2.49988 2.00000 1.00000
dR_dt  <- RR
> as.vector(dR_dt)
[1] 0.5 0.0 0.0
```

This model is scaleable, and would work for any sized disaggregate diffusion model, subject to the available memory resources. The simulation is run with a call to **ode()**, and the output (infected stock) is visualized in Fig. 5.6. The results show the impact of the *CE* matrix values, as the cohort with the highest value (Young) are

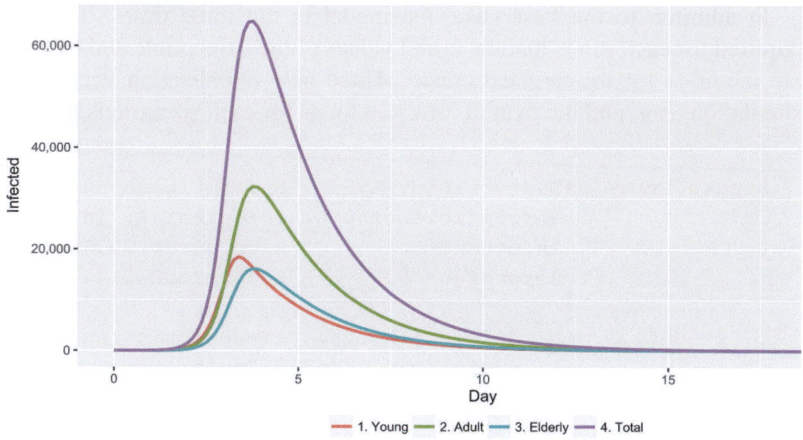

Fig. 5.6 Simulation output from the vectorised SIR model

the first to peak. Additional scenarios concerning what cohort should be vaccinated are now explored.

```
o<-data.frame(ode(y=stocks, times=simtime, func = model,
                  parms=auxs, method="euler"))
```

Policy Exploration with the Disaggregate SIR Model

This R model can support policy analysis, and the following scenario is proposed.

> Assume that a new strain of influenza is circulating, and that for the population of 100,000 there is only a stockpile of 20,000 vaccines. The goal is to investigate whether targeting specific cohorts with the vaccine will make a difference to the overall outcome.

The four scenarios are summarized in Table 5.4, and these include a base case where no vaccines are administered. Furthermore, no logistical difficulties are assumed, such as transportation delays of the vaccine to locations, or capacity constraints in frontline health care services.

Table 5.4 Scenarios exploring vaccination policies

Scenarios	Recovered young	Recovered adult	Recovered elderly
(1) No interventions	0	0	0
(2) Vaccinate young	20,000	0	0
(3) Vaccinate adult	0	20,000	0
(4) Vaccinate elderly	0	0	20,000

In addition to the base case, the model is run three times. The only change required for each run is that the initial values of the susceptible and recovered stocks are modified for the targeted cohort. Three new initialization vectors are created, simulations ran, and the overall infection totals for each scenario is then aggregated.

```
stocks.vy <- c(SusceptibleY=4999,  SusceptibleA=50000,
               SusceptibleE=25000, InfectedY=1, InfectedA=0,
               InfectedE=0,        RecoveredY=20000,
               RecoveredA=0,       RecoveredE=0)

o.vy<-data.frame(ode(y=stocks.vy, times=simtime,func = model,
                  parms=auxs, method="euler"))

stocks.va <- c(SusceptibleY=24999, SusceptibleA=30000,
               SusceptibleE=25000, InfectedY=1,InfectedA=0,
               InfectedE=0,        RecoveredY=0,
               RecoveredA=20000,   RecoveredE=0)

o.va<-data.frame(ode(y=stocks.va, times=simtime,func = model,
                     parms=auxs, method="euler"))

stocks.ve <- c(SusceptibleY=24999, SusceptibleA=50000,
               SusceptibleE=5000,  InfectedY=1, InfectedA=0,
               InfectedE=0,        RecoveredY=0,
               RecoveredA=0,       RecoveredE=20000)

o.ve<-data.frame(ode(y=stocks.ve, times=simtime,func = model,
                     parms=auxs, method="euler")
```

The simulation output is shown in Fig. 5.7, with the total numbers infected for each scenario plotted. It highlights a difference in infection dynamics depending on which cohort is vaccinated. The variation arises on two fronts:

- The peak value of the infected curve differs significantly over the four scenarios. With no vaccination, the peak is at its highest, which would be expected. The next highest peak is for the elderly cohort, closely followed by the adult cohort. By far the lowest peak is obtained when the young cohort is vaccinated. The reason for this due to effective contact values, which are higher in the young when compared to the other cohorts. This confirms that targeting cohorts with the highest effective contact rates can reduce the peak of the curve, which can have a practical benefit in terms of reducing stresses on health systems infrastructure.

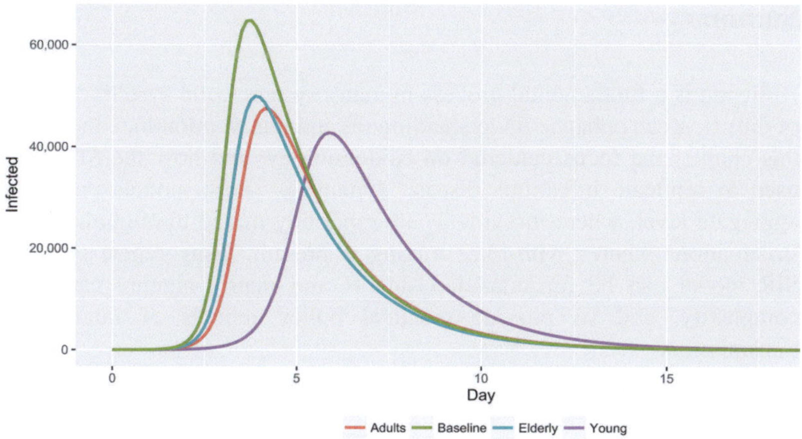

Fig. 5.7 The impact of targeting vaccines to cohorts

- The time taken to reach the peak also varies depending on which cohort is targeted for vaccination. In these simulation runs the peak time for the young cohort occurs latest in the simulation. This shows that selective targeting of vaccines to cohorts with the highest effective contact rates can slow down the pace of contagion. Slowing down the spread provides public health officials with additional time to implement other containment strategies, such as reducing social contacts.

The advantage of the disaggregate SIR model is that also facilitates more detailed and realistic analysis of heterogeneous social mixing, and how that impacts the spread of a virus. Furthermore, it also provides the scope to assess the impact of social distancing measures. For instance, analyzing the impact of school closures is now possible, as this would involve applying an effect variable for the parameter β_{YY}, as shown in Eq. (5.67), and running the simulation.

$$\beta_{YY} = \beta'_{YY} \times \textit{Effect of School Closures on } \beta_{YY} \tag{5.67}$$

In this case β'_{YY} is the reference value, and a case study by Jackson et al. (2011) showed that a school closures was associated with a 65 % reduction in the mean total number of contacts for each student. This information could be added to the model to support further scenario analysis and policy design.

Summary

Diffusion is a fundamental process in many systems, and system dynamics models of diffusion can enhance understanding of, and intervention in, complex systems. In this chapter the focus centered on epidemiology, and how the SIR model can be used to replicate infectious disease dynamics. These models can operate at an aggregate level, where individuals are randomly mixed throughout the population. In situations where "with-like" mixing is present, disaggregate system dynamics SIR model can be formulated. Using R and matrix algebra can reduce model complexity, and so provide practical policy models of inter-cohort disease transmission.

Exercises

1. Suppose we have a town with 10,000 (=N) individuals, of which 1 % were infectious with measles, with $R_0 = 12$ and $D = 7$ days.
   - Calculate the force of infection λ
2. Specify a stock and flow model to simulate the spread of influenza. Assume that the value for R_0 is 2, and that the average recovery delay is 2 days. The model should have the following features:
   - Its core structure should be based on the Susceptible—Infected—Recovered model.
   - It should cater for three policy options. First, it should allow for vaccinations, through a vaccination fraction V_F. Second, it should allow for quarantine, though a quarantine fraction Q_F. Finally, it should model social distancing measures such a school closures, by providing a damping coefficient on C_D on the value of R_0.
   - Assume the following values for these constants: $V_F = 0.15$, $Q_F = 0.08$ and $C_D = 0.81$.
   - All of the policy options should be activated/deactivated through the use of a control flag. Each flag has a value of 0 (policy deactivated) or 1 (policy activated).
3. Draw a stock and flow model (with equations), based on the following set of differential equations that model the Susceptible-Exposed-Infected-Recovered model.

$\frac{dS}{dt} = -\lambda S$	$\frac{dE}{dt} = \lambda S - fE$
$\frac{dI}{dt} = fE - rI$	$\frac{dR}{dt} = rI$

References

Anderson R, May R (1992) Infectious diseases of humans. Oxford University Press, Oxford

Borgdorff MW, Nagelkerke NJ, Broekmans JF (1999) Transmission of tuberculosis between people of different ages in The Netherlands: an analysis using DNA fingerprinting. Int J Tuberc Lung Dis 3(3):202–206

Breban R, Vardavas R, Blower S (2007) Theory versus data: how to calculate R_0? PLoS ONE 2 (3): e282 (PMC, Baylis M (ed))

Dangerfield BC, Fang Y, Roberts CA (2001) Model-based scenarios for the epidemiology of HIV/AIDS: the consequences of highly active antiretroviral therapy. Syst Dyn Rev 17(2):119–150

Jackson C, Mangtani P, Vynnycky E, Fielding K, Kitching A, Mohamed H, Maguire H (2011) School closures and student contact patterns. Emerg Infect Dis 17(2):245

Glass LM, Glass RJ (2008) Social contact networks for the spread of pandemic influenza in children and teenagers. BMC Public Health 8(1):61

Mossong J, Hens N, Jit M, Beutels P, Auranen K, Mikolajczyk R, Edmunds WJ (2008) Social contacts and mixing patterns relevant to the spread of infectious diseases. PLoS medicine 5(3): e74

Thompson KM, Tebbens RJD (2008) Using system dynamics to develop policies that matter: global management of poliomyelitis and beyond. Syst Dyn Rev 24(4):433–449

Vynnycky E, White R (2010). An introduction to infectious disease modelling. Oxford University Press, Oxford

Chapter 6
Model Testing

Model testing should be designed to uncover errors so that you and your clients can understand the model's limitations, improve it, and ultimately use the best available model to assist in important decisions.

John D. Sterman, Business Dynamics (2000, p. 846).

Abstract This chapter provides an overview of model testing in system dynamics, and presents practical methods—using the R framework—that can be used to develop automated model tests. An important challenge in system dynamics is to build client confidence in models. While there is no single test that serves to validate a system dynamics model, confidence in a model gradually accumulates as the model passes more tests. Testing should not be designed to prove that a model is right, as all models are simplified representations of the world. However, models can be useful, and performing a wide range of tests on models can uncover errors. The chapter shows how R can be used to support automated testing of system dynamics models, and also how the concept of the atomic behavior pattern can support behavior tests.

Keywords Validation · Automated tests · RUnit · Behavior modes · Mutation testing

Model Validation in System Dynamics

Models can be classified in different ways, and for the purposes of validation, a distinction can be made between models that are *causal-descriptive* and models that are *correlational* (Barlas 1996). Causal-descriptive models are *white-box* and present a causal theory that can reproduce the system's behavior, and also explain how the behavior is generated. System dynamics models are *causal-descriptive*. Correlational models are *black-box* and data-driven, where there is no claim of causality in the model structure, and these models are validated if the output matches the data within a specified range. Regression and time-series models are

J. Duggan, *System Dynamics Modeling with R*, Lecture Notes in Social Networks,
DOI 10.1007/978-3-319-34043-2_6

correlational. The same dynamic problem can be explored using both model types. For example, in Chap. 5, the SIR model of disease transmission was presented, and this is a *white-box* method to explore policy responses to infectious disease outbreaks. Disease dynamics can also be predicted using time-series *black-box* modeling (Viboud et al. 2003), which used historical data to predict future outbreaks, using time-series forecasting algorithms. Given that system dynamics models are *causal-descriptive*, the ultimate objective of model validation in system dynamics is to (1) establish the model's structural validity, and (2) to evaluate the model's behavioral validity (Barlas 1996).

Structural validity is assessed through comparisons with knowledge of the real-world system structure (Barlas 1996), and examples of structural tests are now summarized.

- *Structure confirmation test*, which is an empirical test that questions whether the model structure is consistent with the real-world system. To pass a structure confirmation test, the model structure must not contradict knowledge about real world system structure (Forrester and Senge 1980). Techniques that can be deployed for these tests include stock and flow maps, direct inspection of model equations, and workshops to gather expert opinion (Sterman 2000). These tests are practical and can be conducted when exploring model structure with end-users and domain experts. For example, in the health systems model of Chap. 4, domain experts such as general practitioners, who would be involved in the model building process, could also provide feedback as to whether the stocks and flows in the model adequately captured the causal structure of the real-world system. No doubt that such a dialogue would highlight that a missing stock in the general practitioner model is the supply line of graduates for the medical profession. Therefore, it is likely that this initial model would not pass a full structure confirmation test.
- *Parameter confirmation*, which involves evaluating parameters against knowledge of the real system, both *conceptually* and *numerically* (Forrester and Senge 1980). Conceptual correspondence means that parameters align with the system structure. For example, in the SIR model from Chap. 5 the recovery delay parameter can be mapped onto the actual physiological process where it takes time for infectious people to recover. Numerical confirmation involves determining if the value of the parameter falls within a plausible range, as it is crucial that system dynamics models strive to describe real decision-making processes (Forrester and Senge 1980). In this case, a parameter such as effective contact rate (C_E) would have to be within plausible boundaries that would make sense to epidemiologists.

- *Direct extreme-condition testing*, which involves evaluating the validity of model equations under extreme conditions. In constructing an extreme-conditions test, the rate (policy) equations in the model are examined, and traced back to the stocks that the rate depends on (Forrester and Senge 1980). The implications of extreme values (zero, minus infinity, plus infinity) can be assessed, and the model tested for these conditions. Extreme-condition tests are powerful tests for discovering flaws in model structure, for example, examining conformance to basic physical laws. For example, the health systems model defined in Chap. 4 could have tests to ensure that no completed visits can occur if there are no general practitioners in the system, or if the productivity of general practitioners was set to zero.
- *Dimensional consistency*, which involves dimensional analysis of the model's equations, and has been discussed in Chap. 1. Every equation in a system dynamics model should be dimensionally consistent, and many system dynamics tools have a feature to automatically check for this. A persuasive reason for using dimensional analysis is that failure to pass this test often reveals faults in the underlying model structure (Forrester and Senge 1980).
- *Boundary adequacy*, where the purpose is to assess whether the important concepts for addressing the problem are endogenous to the model (Sterman 2000), and whether the model behavior significantly changes as the boundary is extended.

Once confidence in the structural properties of the model has been established, *behavioral validity* tests are can be used to assess how well the model output aligns with the observed real-world behavior. Barlas (1996) presents a procedure for behavior pattern validation, which classifies the approach based on two different behavior patterns.

- For problems that involve transient highly non-stationary behavior, an approach is to compare graphical-oriented measures such as: amplitude of peak, time between two peaks, minimum value, slope and the number of inflection points.
- For steady state simulation results, a six-step statistical procedure, proposed by Barlas (1989), can be utilized. For this, there are two sets of observations, the simulated data $S = (S_1, S_2, \ldots, S_N)$ and the actual time-series data $A = (A_1, A_2, \ldots, A_N)$.

The appropriate mathematical equations for the steady state scenario are provided in Table 6.1. To utilize these behavior tests for steady state simulation data, both time-series data (S), and simulation output (A) are required. A exemplar synthetic model, based on epidemic dynamics with a dynamic population (births and deaths), is presented by Barlas (1989). This provides the interested reader with an excellent basis to further explore this behavioral technique. The elements of the six-step process are now summarized.

1. *Trend comparison and removal.* Because most statistical procedures require stationarity in the means, trend comparison and removal is required. In time series analysis, a stationary process is one whose statistical properties such as

Table 6.1 Selected equations for the multi-step validation procedure (Barlas 1996)

Trend comparison and removal (linear)	$\hat{Y} = b_0 + b_1 t$ $Z_i = Y_i + \hat{Y}_I$
Autocovariance function of a time-series for lag $k = 0, 1, 2, \ldots < N$	$Cov(k) = \frac{1}{N} \sum_{i=1}^{N-k} (x_i - \bar{x})(x_{i+k} - \bar{x})$
Autocorrelation function of a time-series for lag $k = 0, 1, 2, \ldots < N$	$r(k) = \frac{Cov(k)}{Cov(0)} = \frac{Cov(k)}{Var(X_i)}$
Comparing percent error in means (E_1) and variations (E_2) for S and A	$E_1 = \frac{\lvert\bar{S}-\bar{A}\rvert}{\bar{A}}, \quad E_2 = \frac{\lvert s_S - s_A \rvert}{s_A}$
The cross-correlation function for S and A, $k = 0, 1, 2, \ldots$	$C_{SA}(k) = \frac{\left(\frac{1}{N}\right) \sum_{i=k}^{N} (S_i - \bar{S})(A_{i-k} - \bar{A})}{s_S s_A}$
The cross-correlation function for S and A, $k = 0, -1, -2, \ldots$	$C_{SA}(k) = \frac{\left(\frac{1}{N}\right) \sum_{i=k}^{N} (A_i - \bar{A})(S_{i+k} - \bar{S})}{s_S s_A}$
The discrepancy coefficient U, which ranges from 0 ('perfect predictions') to 1 ('worst predictions')	$U = \frac{\sqrt{\sum (S_i - \bar{S} - A_i - \bar{A})^2}}{\sqrt{\sum (A_i - \bar{A})^2} + \sqrt{\sum (S_i - \bar{S})^2}}$

the mean and variance are constant over time (Cowperthwait and Metcalfe 2009). If there is no significant difference in trends between the two data sets, the trend components can be removed. If there are significant differences in the trends, then that suggests that a model revision is required.

2. *Comparing the periods*. An autocorrelation function test can detect significant errors in the periods, and the test can be used to discover if one behavior pattern has high-frequency components not present in others.
3. *Comparing the means*. When the model has no systematic error, E_1 (see Table 6.1), rarely exceeds 5 %.
4. *Comparing the variations*. Even if the model has no systematic error, this can be as large as 30 %.
5. *Testing for phase lag*. The cross-correlation function provides an estimate of a potential phase lag between the actual and simulated data. In experiments, Barlas (1989) reports that the cross-correlational function quantity $|max - min|$ was always larger than 0.80 in the presence of a systematic phase lag.
6. As a final step, when all other validity tests have passed, a discrepancy coefficient U can be computed as a single summary measure. Models without systematic errors can have U values as high as 0.70, as U is a point prediction measure, whereas system dynamics models are pattern-oriented (Barlas 1989).

R provides excellent support for implementing time-series methods, including special-purpose functions for trend removal, autocovariance and autocorrelation (Cowpertwait and Metcalfe 2009).

Recent advances in toolsets to support behavior analysis and testing software include pattern recognition for model testing, calibration and behavior analysis (Yücel and Barlas 2015), as well as the recent BATS framework (Sücüllü and Yücel 2014), which integrates a pattern classification algorithm and statistical methods to analyze steady-state behaviors. Furthermore, a number of these tests can be quantified and constructed in the form of assertion checking that is widely used in software program verification (Balci 1994), and the R platform can support an automated approach to testing system dynamics models.

Automated Validity Tests

In order to improve model robustness, Peterson and Eberlein (1994) introduced the idea of automated validation testing, and documented a language for specifying tests. For example, in a customer growth model, such as that presented in Chap. 1, the following notation is used to express a test condition, also known as a reality check in Vensim.

```
:THE CONDITION: Customers = 0 :IMPLIES: Recruits = 0
```

This representation allows the modeler to run a simulation where the customer integral is set to zero, and the simulation output is tested to ensure that the corresponding value for recruits is also zero. This can be written as a test condition, which identifies the set of inputs, and the expected output for a given test. Table 6.2 illustrates this, where an individual test has the following properties:

- An identifier (T_i), which uniquely identifies the test.
- The test condition, based on the inputs and expected output, which should evaluate to true after the simulation is finished.
- The set of inputs.
- The expected output.

Based on this approach, the modeler can design test conditions for the simulation model. These can be executed to ensure that the model behaves as expected, and once successful, they can enhance client confidence in the model. A benefit of automated tests, which are widely used in the software development process, is that, once developed, they can be executed at any time, usually following on from a model revision.

Tests can be designed for any system dynamics model, as the initial conditions for a simulation can be set so that actual results can be compared to expected results. The SIR model introduced in Chap. 5 is now revisited in order to explore a

Table 6.2 Test condition representation from customer growth model (Chap. 1)

Test ID	Test condition	Inputs	Expected output
T_1	IF *Customers* (*C*) = 0 THEN *Recruits* = 0	*C* = 0	*Recruits* = 0

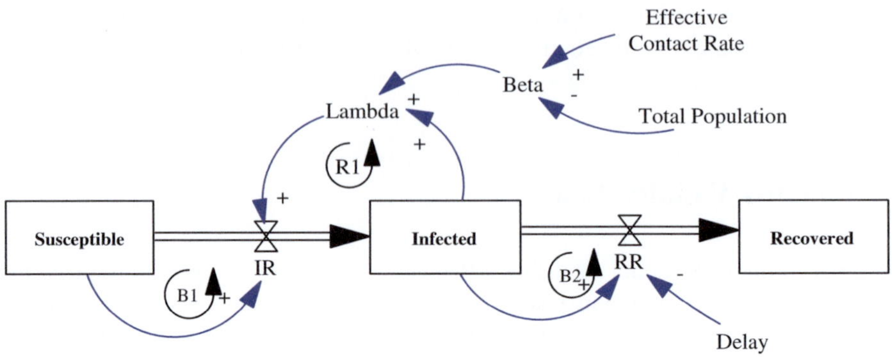

Fig. 6.1 The SIR model revisited

number of test cases, where these can be automated and used to improve model validity. The SIR model is illustrated in Fig. 6.1.

The first set of tests are designed to assess the robustness of the infection rate (*IR*) equation, which is a crucial element of the positive feedback loop. The question to be addressed is under what conditions will the infection rate remain at zero. First, recall the *IR* equation for (6.1), and this is used as a basis for test design.

$$IR = S \times \lambda = S\beta I = S\left(\frac{C_E}{N}\right)I \tag{6.1}$$

Based on Eq. (6.1), there are three scenarios that will ensure that no individual can become infected:

1. With no susceptible people ($S = 0$), there will be no stock of vulnerable individuals to infect.
2. With no infected people ($I = 0$), there are no infected people in circulation that could transmit the virus to susceptible people.
3. With no effective contacts ($C_E = 0$), there are no contacts in the population, therefore transmission cannot occur.

Based on these scenarios, three tests can be specified, and these are shown in Table 6.3. The inputs include the initial values of the variables for a simulation run (the three stocks and the effective contact rate), and the combination of these values that should generate the expected result in each case.

The fourth test (T_4) focuses on the recovery rate, which is the outflow from the infected stock. In this scenario, the test is to explore the conditions that would result in a recovery rate (*RR*) of zero. One way to achieve this is to make the infected stock zero, as with T_2. The second way is to design a *loop knockout test* (Sterman 2000), by setting the delay constant on the outflow to infinity (∞), which has the effect of deactivating the negative feedback loop (B_2). Table 6.4 specifies this loop knockout test.

Table 6.3 Test conditions for evaluating the infection rate (IR)

Test ID	Test condition	Inputs	Expected output
T_1	IF *Susceptible* (*S*) = 0 THEN $IR = 0$	$S = 0, I = 10000$ $R = 0, C_E = 2$	$IR = 0$
T_2	IF $I = 0$ THEN $IR = 0$	$S = 10000, I = 0, R = 0,$ $C_E = 2$	$IR = 0$
T_3	IF $C_E = 0$ THEN $IR = 0$	$S = 9999, I = 1$ $R = 0, C_E = 0$	$IR = 0$

Table 6.4 Test conditions for evaluating the infection rate (IR)

Test ID	Test condition	Inputs	Expected output
T_4	IF $D = \infty$ THEN RR = 0	$S = 0, I = 10000$ $R = 0, C_E = 2, D = \infty$	$RR = 0$

Table 6.5 Test conditions for evaluating the infection rate (IR)

Test ID	Test condition	Inputs	Expected output
T_5	$\{S, I, R \geq 0\}, \{IR, RR\} \geq 0, \lambda \geq 0$	$S = 9999$ $I = 1, R = 0$ $C_E = 20$	$\{S, I, R\} \geq 0$ $\{IR, RR\} \geq 0$ $\lambda \geq 0$

A test is required to ensure that model variables operate with valid ranges. Clearly, knowledge of the domain will inform this analysis, and in this case, for modeling infectious diseases, there can be no negative values in the model. Therefore test T_5 will test that all stocks, flows and auxiliaries are zero or greater. This condition is specified in Table 6.5.

To date, the comparisons of actual to expected in the tests have involved comparing numeric values. However, a further model test is useful, and this is known as a behavior pattern test. This is based on an attribute of simulation output known as the *atomic behavior pattern* (Ford 1999), and can be viewed as the essential possible shapes of dynamics behavior. This measure can have three possible values for a given variable *x*:

- exponential atomic behavior pattern

$$\frac{\partial\left(\left|\frac{\partial x}{\partial t}\right|\right)}{\partial t} > 0 \tag{6.2}$$

- logarithmic atomic behavior pattern

$$\frac{\partial\left(\left|\frac{\partial x}{\partial t}\right|\right)}{\partial t} < 0 \tag{6.3}$$

- linear atomic behavior pattern

$$\frac{\partial\left(\left|\frac{\partial x}{\partial t}\right|\right)}{\partial t} = 0 \tag{6.4}$$

For these calculations, the net rate of change of the variable of interest is $\partial x/\partial t$, Where the variable is a stock, the net rate of change is simply the net flow, which is readily available in all system dynamics models. The absolute value of this is calculated, and then the derivative of this absolute value with respect to time describes the movement of the net rate of change. As described earlier with the three atomic behavior mode equations, this movement can be described in three ways.

- When the value greater than zero, the atomic behavior pattern is *exponential* (6.2).
- When the value less than zero, the atomic behavior pattern is *logarithmic* (6.3).
- When the value equals zero, the atomic behavior pattern is linear (6.4).

Many complex systems follow atomic behavior patterns. For example, the spread of a virus, as measured through the numbers infected, can be decomposed into a sequence of atomic behavior patterns. The model output of an epidemic scenario (i.e. the value for R_0 is greater than 1) is shown in Fig. 6.2. Application of Eqs. (6.2)–(6.4) on the data set yields interesting observations, and a clear pattern of behavior, as indicated by the colors on the graph.

The curve's behavior is initially exponential (red), as the second derivative is greater than zero. Then, following a point of inflection between time 5 and 7.5, the behavior changes to logarithmic (blue). Once the curve peaks, it declines initially at an exponential rate (red), before leveling off with a logarithmic pattern (blue). This

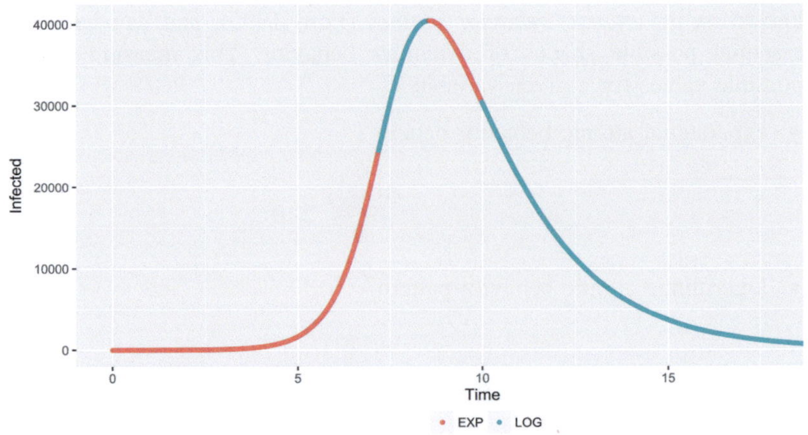

Fig. 6.2 Atomic behavior pattern for the infected variable from the SIR model

information can then be codified and used as part of the testing process, in order to ensure that the actual model behavior adheres to expectations.

The logic for calculating the three atomic behavior modes, and expressing these in a compact form, can be coded in R, in the form of two new functions. The function **bmode()** implements (6.2–6.4), as it accepts the initial net flow, and simulation time, as vectors, and returns the relevant behavior mode as a string vector.

```
bmode<-function(net_flow, time)
{
  net_flow<-abs(net_flow)
  net_flow_dt<- rep(NA, length(net_flow))
  net_flow_dt[2:length(net_flow_dt)]<-diff(net_flow)/
                                      diff(time)
  bm<-ifelse(net_flow_dt > 0,ans<-"EXP",
             ifelse(net_flow_dt < 0,ans<-"LOG",ans<-"LIN"))
  bm[1] = bm[2]
  return(bm)
}
```

The R function **rep()** is used to allocate memory for the result, which will be the length of the net flow vector. The derivative of the net flow is obtained using R's **diff()** function, which returns the difference between successive vector elements, divided by the difference in simulation times. The vectorized **ifelse()** is utilized to classify the modes, based on the second derivative's value. While this function returns a vector containing the behavior mode for each time step, what is important for testing purposes is to identify the correct sequence of atomic behavior modes. In order to achieve this, an additional function is used to extract the reduced form of the behavior pattern. This function is named **bpattern()**.

```
bpattern<-function(bm)
{
  return (rle(bm)$values)
}
```

The function **bpattern** uses R's **rle()** function (run length encoding) to compress the sequence of behavior modes, and so remove any repeating values. This function returns a list of two elements, where the first element contains a vector of the lengths for each element (information that is not used), and the second vector contains the sequence of elements. Therefore, it is this second list element that is returned from the function.

Table 6.6 Test conditions for evaluating the behavior mode

Test ID	Test condition	Inputs	Expected output
T_6	Mode(Infected) = {"EXP", "LOG", "EXP", "LOG"}	$S = 9999$ $I = 1$, $R = 0$ $C_E = 2$	{"EXP", "LOG", "EXP", "LOG"}

```
> rle(t$mode)
Run Length Encoding
  lengths: int [1:4] 565 136 144 1156
  values : chr [1:4] "EXP" "LOG" "EXP" "LOG"
```

As an illustration, the **bpattern** function returns the following for the simulation shown in Fig. 6.2.

```
> bpattern(bmode(t$IR-t$RR,t$time))
[1] "EXP" "LOG" "EXP" "LOG"
```

Therefore, this compressed vector provides the desired behavior mode for a simulation run where R_0 is greater than 1, and where an infected person is introduced into a totally susceptible population. This is useful, as it now provides the necessary information to formalize a behavior mode test, and this test (T_6) is specified in Table 6.6.

Six tests are now designed and they can be applied to the SIR model. The challenge is to find an efficient way to write, execute and analyze the test output. Automating the process is highly desirable, as this allows for a continuous test process, whereby once model changes are made, a full set of tests can then be executed. As a software development environment, R includes a unit testing framework—*RUnit*—which can be deployed to streamline the testing process.

Test Automation with RUnit

Test automation is an important component of modern software development, as manual test methods are insufficient to support an environment of daily software builds (Lo Giudice 2013). In system dynamics, a similar environment of continuous revision and release of models exists, and therefore test automation can play a role in improving model reliability, and hence maximizing client confidence. A schematic of the test automation process is shown in Fig. 6.3. This involves creating a set of tests (T_1, T_2,…, T_N), known as a *test suite*, and executing these in sequence. Each test has an expected outcome for a given set of inputs, and this outcome is evaluated against the actual result. If the individual test passes, the next

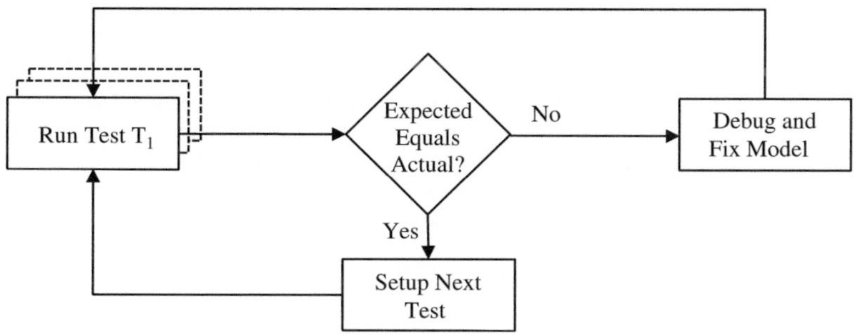

Fig. 6.3 The test automation cycle

test is executed, otherwise the model is debugged and fixed, before resuming the test process once more.

The package **RUnit** (König et al. 2015) provides a convenient structure to design and implement automated tests. Specifically, it provides three supporting R functions that can be used to develop a suite of tests for any system dynamics model. They are:

- **defineTestSuit()**, which creates a test suite, and includes details on the path to the test files, a pattern to match test files, and a pattern to match test functions. The pattern matching approach supports easy extension of tests, as the framework searches through folders and files to automatically find individual tests.
- **isValidTestSuite()**, which validates any given test suite before it is executed, to ensure that the files are properly referenced.
- **runTestSuite()**, which is the central function of the **RUnit** package. It identifies and opens the test files, and executes all matching test functions.

RUnit also provides a set of functions that can be used to test for error conditions, and each of these will evaluate to either TRUE or FALSE. The results are automatically collated by **RUnit**. These functions are listed in Table 6.7, and cater for a range of test conditions where two variables are being compared.

In order to setup the automated process, the set of R files need to be organized in a certain way, and this overall structure is shown in Fig. 6.4.

There are three R files created to facilitate the automated test process:

- *SIR Model.R*, which contains the system dynamics model that needs to be validated.
- *TestSuite.R*, which contains all of the test functions for the model, and in this example, these will be an implementation of the tests $T_1,\ldots, T_6$.
- *TestRunner.R*, which contains a brief script to orchestrate the tests, and will create, validate and execute all tests, before displaying the results.

Table 6.7 Range of RUnit check functions (König et al. 2015)

RUnit function	Description
checkEquals(o1,o2)	Compares two R objects by invoking R's **all.equal()** function on the two objects. If the objects are not equal an error is generated and the failure is reported to the test logger such that it appears in the test protocol. Additional parameters can be provided, including a message, and the tolerance used for comparison, which is useful when comparing floating point numbers
checkEqualsNumeric (o1,o2)	Operates similar to checkEquals except that it invokes R's **all.equal. numeric()** function instead of **all.equal()**
checkIdentical(o1, o2)	A wrapper around R's **identical()** function which uses the error logging mechanism of RUnit
checkTrue(expr)	Checks if the expression provided as first argument evaluates to TRUE. If not, an error is generated and the failure is reported
DEACTIVATED (msg)	Interrupts the test function and reports the test case as deactivated. In the test protocol deactivated test functions are listed separately

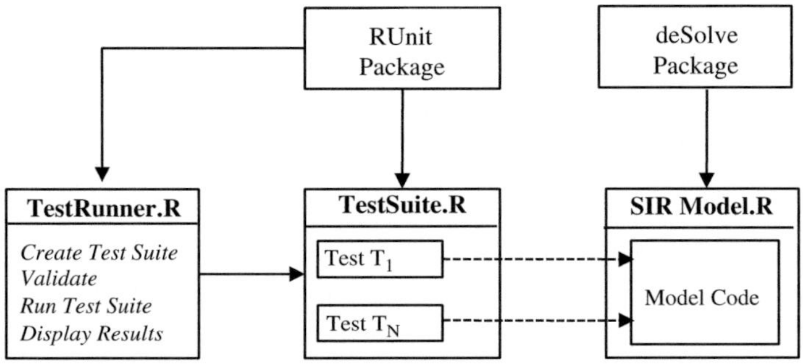

Fig. 6.4 File structure for organizing automated tests

The SIR model implementation is shown, and is similar to previous models except that the simulation time, auxiliaries and stocks are not defined in this source file. This provides additional flexibility for calling the model under a range of different initial conditions, which is a requirement for implementing a set of tests.

```
library(deSolve)

model <- function(time, stocks, auxs){
  with(as.list(c(stocks, auxs)),{
    aBeta     <- aCE / aTotalPopulation
    aLambda   <- aBeta * sInfected

    fIR       <- sSusceptible * aLambda
    fRR       <- sInfected / aDelay

    dS_dt     <- -fIR
    dI_dt     <- fIR - fRR
    dR_dt     <- fRR

    return (list(c(dS_dt,dI_dt,dR_dt),IR=fIR,RR=fRR,
                       Beta=aBeta,Lambda=aLambda,CE=aCE))
  })
}
```

The user-defined R functions that implement the tests described in Tables 6.3, 6.4, 6.5 and 6.6 are now specified. The function's name is based on the test objective, and includes information on the test number. All test functions share a similar naming convention as they begin with the letter "T". The general approach used in each test is as follows:

1. Set the start time, finish time, and simulation time step.
2. Create the simulation time vector.
3. Create the vector of stocks, along with the initial values.
4. Create the vector of auxiliaries, along with the initial values.
5. Call the simulation model via the **ode()** function, and store the simulation output in a data frame.
6. Add a column to the output data frame which contains the expected result from the simulation.
7. Call the appropriate RUnit method to check if the result is as expected, where these methods are selected from the available set listed in Table 6.7.

The first function tests to ensure that the infection rate (*IR*) is zero when there are no susceptible individuals in the population. As with the general approach just defined, the initial conditions are specified in the **stocks** and **auxs** variables. The simulation results are stored in the data frame **t**, and a new column is added (**t $Expected**) with all its values set to zero. RUnit's **checkEquals()** function performs an element-wise comparison—for every time step—on the two data frame columns. The RUnit framework records the result of the test, and a call can be made to display this once all the tests are completed.

```
T01.S.Eq.0.IMP.IR.Eq.0<-function()
{
  START<-0; FINISH<-20; STEP<-0.01;
  simtime <- seq(START, FINISH, by=STEP)
  stocks  <- c(sSusceptible=0,sInfected=10000,sRecovered=0)
  auxs    <- c(aTotalPopulation=10000, aCE=2, aDelay=2)
  t<-data.frame(ode(y=stocks, times=simtime, func = model,
                  parms=auxs, method="euler"))
  t$Expected<-0
  checkEquals(t$Expected,t$IR)
}
```

The second and third test functions follow a similar pattern. These functions also begin with the letter "T" and their names reflect the test's success condition. As with the first function, the expected results are added as a column to the data frame, and these are compared to the actual values using the **checkEquals()** function.

```
T02.I.Eq.0.IMP.IR.Eq.0<-function()
{
  START<-0; FINISH<-20; STEP<-0.01;
  simtime <- seq(START, FINISH, by=STEP)
  stocks  <- c(sSusceptible=10000,sInfected=0,sRecovered=0)
  auxs    <- c(aTotalPopulation=10000, aCE=0, aDelay=2)
  t<-data.frame(ode(y=stocks, times=simtime, func = model,
                  parms=auxs, method="euler"))
  t$Expected<-0
  checkEquals(t$Expected,t$IR)
}
```

```
T03.CE.Eq.0.IMP.IR.Eq.0<-function()
{
  START<-0; FINISH<-20; STEP<-0.01;
  simtime <- seq(START, FINISH, by=STEP)
  stocks  <- c(sSusceptible=9999,sInfected=1,sRecovered=0)
  auxs    <- c(aTotalPopulation=10000, aCE=0, aDelay=2)
  t<-data.frame(ode(y=stocks, times=simtime, func = model,
                  parms=auxs, method="euler"))
  t$Expected<-0
  checkEquals(t$Expected,t$IR)
}
```

The fourth test, which focuses on the recovery rate, follows a similar pattern to the first three tests, and utilizes R's **Inf** value, which is a built-in value that represents a value for infinity. This can be used in any equation, for example, 1 divided by **Inf** returns a value of 0.

```
T04.D.Eq.INF.IMP.RR.Eq.0<-function()
{
  START<-0; FINISH<-20; STEP<-0.01;
  simtime <- seq(START, FINISH, by=STEP)
  stocks  <- c(sSusceptible=0,sInfected=10000,sRecovered=0)
  auxs    <- c(aTotalPopulation=10000, aCE=2, aDelay=Inf)
  t<-data.frame(ode(y=stocks, times=simtime, func = model,
                parms=auxs, method="euler"))
  t$Expected<-0
  checkEquals(t$Expected,t$RR)
}
```

The fifth test checks for any negative values in the model's variables by using the **checkTrue()** function, based on calls to R's **all()** function.

```
T05.IMP.All.Vars.GTE.0<-function()
{
  START<-0; FINISH<-20; STEP<-0.01;
  simtime <- seq(START, FINISH, by=STEP)
  stocks  <- c(sSusceptible=9999,sInfected=1,sRecovered=0)
  auxs    <- c(aTotalPopulation=10000, aCE=20, aDelay=2)
  t<-data.frame(ode(y=stocks, times=simtime, func = model,
                    parms=auxs, method="euler"))
  checkTrue(all(t$sSusceptible>=0))
  checkTrue(all(t$sInfected >= 0))
  checkTrue(all(t$sRecovered>=0)); checkTrue(all(t$IR>=0))
  checkTrue(all(t$RR>=0)); checkTrue(all(t$Lambda>=0))
}
```

The **all()** R function is a powerful way to apply the same test to every element of a vector, and convenient for testing that all simulated values for the model's stocks, flows and auxiliaries are positive.

The final test evaluates the behavior pattern that follows a "boom and bust" dynamic, which is the classic trajectory for infectious diseases. This test is feasible because of the earlier defined functions, **bmode()** and **bpattern()**. The expected output is stored in the string vector **expected**, and this is compared to the actual result calculated, using the function **checkEquals()**.

```
T06.IMP.Mode.BellShaped<-function()
{
  START<-0; FINISH<-20; STEP<-0.01;
  simtime <- seq(START, FINISH, by=STEP)
  stocks  <- c(sSusceptible=9999,sInfected=1,sRecovered=0)
  auxs    <- c(aTotalPopulation=10000, aCE=2, aDelay=2)
  t<-data.frame(ode(y=stocks, times=simtime, func = model,
                          parms=auxs, method="euler"))
  expected<-c("EXP","LOG","EXP","LOG")
  actual<-bpattern(bmode(t$IR-t$RR,t$time))
  checkEquals(expected,actual)
}
```

Once all the test functions are written, the final step in the test automation process is to implement the *TestRunner.R* file, which controls the test automation sequence. Because the process uses R's pattern matching utilities, this file is short, and contains the minimum number of statements to setup all the tests.

```
# file TestRunner.R
library(RUnit)

testsuite.SIR<-defineTestSuite("SIR Model Tests",
         dirs=file.path("models/06 chapter/R/tests/"),
         testFileRegexp = "^TestSuite.+\\.R",
         testFuncRegexp = "^T.+")

if(isValidTestSuite(testsuite.SIR))
{
  test.result<-runTestSuite(testsuite.SIR,verbose=0)
  summary(test.result)
}
```

The function **defineTestSuite()** is called first, and this makes use of *regular expressions* to find the correct files and functions to process. A regular expression is a sequence of characters that define a search pattern, and they are mainly used for pattern matching. The parameters passed to **defineTestSuite()** are:

- The test suite name, which should be informative, as a large project could have many test suites.
- The path to the directory location of all the test files, which is usually a sub-folder in the model file's directory.
- A regular expression (parameter *testFileRegexp*) that contains a pattern for finding test files. In this case, all R files beginning with the text "TestSuite" and ending in ".R" will be identified as test suite scripts.

Table 6.8 Regular expression symbols in R

Symbol	Description
.	Matches any character
^	An anchor that matches the start of a string
$	An anchor that matches the end of a string
+	The preceding item will be matched one or more times
\\	Escape a special character

- A regular expression (parameter *testFuncRegexp*) that contains a pattern for finding test function within each file. In this case, all R functions beginning with "T" will be executed.

For regular expressions, the idea is that a *pattern string* is created that specifies a search rule that can be used by special R functions. An example of one search function is **grep(pattern, x)**, where *pattern* is a regular expression, and *x* is the vector to be searched. This function returns an index of matching value locations. A selection of special symbols can be used to define the search, and a number of these are summarized in Table 6.8.

To illustrate this, and gain an insight into how the RUnit framework operates, consider the following vector, which contains a list of file names.

```
> t<-c("TestSuite1.R","TestSuite.C","TestSuite2.R")
> t
[1] "TestSuite1.R" "TestSuite.C"  "TestSuite2.R"
```

The goal is to find all files starting with "TestSuite" and ending in ".R". A pattern string ***p*** must be created to specify the search rule, and this is shown below.

```
> p<-"^TestSuite.+\\.R"
> p
[1] "^TestSuite.+\\.R"
```

While this may appear somewhat cryptic, the rule specifies the following checks.

- The first 9 characters of the target vector element must equal "TestSuite"
- After that, any number of characters are matched until the string ".R" is reached.
- The escape characters \\ are needed because the "dot" in ".R" is itself a special regular expression symbol.

This pattern is then passed as a parameter to the R function **grep()**.

```
> t
[1] "TestSuite1.R" "TestSuite.C"  "TestSuite2.R"
> grep(p,t)
[1] 1 3
```

It is interesting to note that the returned vector contains the indices of the two matching vector elements. This vector can then be applied to the original vector to filter the results, which clearly show that only the two R files have been selected.

```
> t[grep(p,t)]
[1] "TestSuite1.R" "TestSuite2.R"
```

The use of regular expressions provides excellent scalability for tests. It means that once the correct naming convention is used, the *TestRunner.R* file will automatically detect any number of files and tests that have been specified. Sample output from defining the test suite is shown below.

```
> testsuite.SIR<-defineTestSuite("SIR Model Tests",
+       dirs=file.path("models/06 chapter/R/tests/"),
+       testFileRegexp = "^TestSuite.+\\.R",
+       testFuncRegexp = "^T.+")

> str(testsuite.SIR)
List of 6
 $ name           : chr "SIR Model Tests"
 $ dirs           : chr "models/06 chapter/R/tests/"
 $ testFileRegexp: chr "^TestSuite.+\\.R"
 $ testFuncRegexp: chr "^T.+"
 $ rngKind        : chr "Marsaglia-Multicarry"
 $ rngNormalKind : chr "Kinderman-Ramage"
 -  attr(*, "class")= chr "RUnitTestSuite"
```

In addition to recording the initial parameters, the list also shows the default values for two additional attributes: **rngKind** and **rngNormalKind**, which refer to the default random number generator algorithms. The next step is to validate the test script by calling the function **isValidTestSuite()**, which will check that the directory path is valid. In this case, the paths are valid, and the call returns TRUE.

```
> str(isValidTestSuite(testsuite.SIR))
 logi TRUE
```

Once it is valid, the function **runTestSuite()** is invoked, and this returns a nested list. R's **summary()** function summarizes this output in a user-friendly manner.

```
> summary(test.result)
RUNIT TEST PROTOCOL
***********************************************
Number of test functions: 6
Number of errors: 0
Number of failures: 0

1 Test Suite :
SIR Model Tests - 6 test functions, 0 errors, 0 failures

Details
***************************
Test Suite: SIR Model Tests
Test function regexp: ^T.+
Test file regexp: ^TestSuite.+\.R
Involved directory:
models/06 chapter/R/tests/
---------------------------
Test file: models/06 chapter/R/tests//TestSuite.SIR.R
T01.S.Eq.0.IMP.IR.Eq.0: (1 checks) ... OK (0.11 seconds)
T02.I.Eq.0.IMP.IR.Eq.0: (1 checks) ... OK (0.11 seconds)
T03.CE.Eq.0.IMP.IR.Eq.0: (1 checks) ... OK (0.11 seconds)
T04.D.Eq.INF.IMP.RR.Eq.0: (1 checks) ... OK (0.12 seconds)
T05.IMP.All.Vars.GTE.0: (6 checks) ... OK (0.11 seconds)
T06.IMP.Mode.BellShaped: (1 checks) ... OK (0.18 seconds)
```

In RUnit, a distinction is made between a failure and an error (König et al. 2015). A failure occurs if one of the check functions fail (e.g. checkEquals(0,100) generates a failure). An error is reported if an ordinary R error occurs. For our scenario, all tests pass, based on the initial conditions and the model results.

However, it is also important to assess the quality of the written tests to see how effective they are at detecting error cases. In order to evaluate the efficacy of these tests, a software engineering approach known as *mutation testing* can be used. This involves generating a number of variants of a program (or model), where each of these variants slightly differs from the original version. Variants are based on applying a set of transformations known as mutation operators (Van Vliet 2008). For example, these mutations can be achieved by actions to:

- Replace a constant by another constant
- Replace a variable by another variable
- Replace a constant by a variable
- Replace an arithmetic operator by another arithmetic operator
- Replace a logical operator by another logical operator

Table 6.9 Scenarios for mutating SIR model equations

Scenario	Correct SIR equation	Mutated equation
1	$\beta = \frac{C_E}{N}$	$\beta = C_E \times N$
2	$\lambda = \beta \times I$	$\lambda = \frac{\beta}{I}$
3	$IR = S \times \lambda$	$IR = \frac{\lambda}{S}$
4	$RR = \frac{I}{D}$	$RR = I \times D$

Table 6.10 Summary of test results for mutation scenarios

Scenario	Variable impacted	Tests passed	Tests failed
1	β	{T1, T2, T3, T4}	{T5, T6}
2	λ	{T1, T2, T4, T5}	{T3, T6}
3	*IR*	{T2, T3, T5}	{T1, T4, T6}
4	*RR*	{T1, T2, T3, T5}	{T4, T6}

These actions can now be applied to any system dynamics model, and for the SIR model, four scenarios are identified which replace the arithmetic operator in selected equations (see Table 6.9).

Each of the tests can then be run on these incorrect model equations. For example, the first scenario is run, where the equation for β is mutated, the test framework reports the following failures for T_5 and T_6. The full summary of results is presented in Table 6.10.

```
T05.IMP.All.Vars.GTE.0: FAILURE !! (check number 1)
Error in checkTrue(all(t$sSusceptible >= 0)) : Test not TRUE
T06.IMP.Mode.BellShaped: FAILURE !! (check number 1)
Error in checkEquals(expected, actual) :
  Lengths (4, 1995) differ (string compare on first 4)
'is.NA' value mismatch: 3 in current 0 in target
```

The reasons for the tests failures under these four scenarios are as follows:

- When a fault is injected into β, the two tests that fail are the positivity test for all variables (T_5), and the expected behavior mode (T_6). Interestingly, the first four tests still pass in this scenario, and this can be explained by the fact that the value of β is not crucial for these particular tests.
- When the equation for λ is incorrect (where *I* becomes the denominator) failures are recorded for T_3, due to a divide by zero error for the λ calculation, and the expected behavior mode test (T_6) also fails.
- When *IR* is corrupted, three tests fail. The first test (T_1) fails because the variable *S* is now the denominator for *IR*, and so a divide by zero error occurs. The fourth test (T_4) fails because the stock *I* is set to infinity, which in turn sets the flow *RR* to minus infinity. Because of these effects, the behavior mode test also fails (T_6).

- Finally, when the recovery rate equation *RR* is changed, two tests fail. Test T_4 fails because infinity is no longer a denominator on the equation, and the behavior mode test T_6 also fails because the modified equation does not generate the expected dynamic behaviour pattern.

Subjecting the validity tests to further set of mutation tests provides a means to increase confidence in the test suite that supports the model building process. Overall, the worked SIR model example confirms the benefits of deploying an automated validity test approach to system dynamics models. The list of tests can be easily extended. For example, up to five tests per variable can be written (Peterson and Eberlein 1994), who also recommend that tests should ideally outnumber equations. R's unit test framework provides a scaleable and efficient structure for managing and running a high volume of model tests.

Summary

Model testing is crucial in order to build client confidence in system dynamics models. There are a range of tests that can be conducted to enhance model acceptance. These include tests for structural and behavioral validity. Structural tests are used to confirm that the model stock and flow structure does not contradict knowledge about the real world system. Behavior tests include extreme-condition testing as a method to compare model results to actual data. Established software engineering techniques such as mutation testing can also be used. The R platform supports an automated test process, and special-purpose functions can be written as part of the model building process, in order to perform a suite of automated validity tests.

Exercises

1. Design a set of appropriate tests for the following system dynamics model, originally presented in Chap. 1.

$$Customers = INTEGRAL(Recruits - Losses, 10000)$$
$$Recruits = Customers \times Growth\,Fraction$$
$$Growth\,Fraction = 0.07$$
$$Losses = Customers \times Decline\,Fraction$$
$$Decline\,Fraction = 0.03$$

2. Based on following economic model, specified earlier in Chap. 3, identify the equations that mutation testing could be applied to, and develop an appropriate set of mutation tests.

$$Machines\,(M) = INTEGRAL(Investment - Disards, 100)$$
$$Investment = Economic\ Output \times Reinvestment\ Fraction$$
$$Discards = Machines \times Depreciation\ Fraction$$
$$Reinvestement\ Fraction\,(R) = 0.20$$
$$Depreciation\ Fraction(D) = 0.10$$
$$Economic\ Output\,(O) = Labour \times \sqrt{Machines}$$
$$Labour\ (L) = 100$$

References

Balci O (1994) Validation, verification and testing techniques throughout thelife cycle of a simulation study. Annals of OR 53

Barlas Y (1989) Multiple tests for validation of system dynamics type of simulation models. Eur J Oper Res 42(1):59–87

Barlas Y (1996) Formal aspects of model validity and validation in system dynamics. Syst Dyn Rev 12(3):183–210

Cowpertwait PS, Metcalfe AV (2009) Introductory time series with R. Springer Science & Business Media

Ford, D. N. (1999). A behavioral approach to feedback loop dominance analysis. SystemDynamics Review, 15(1), 3.

Forrester JW, Senge PM (1980) Tests for building confidence in system dynamics models. In: Legasto AA, Forrester JW, Lyneis JM (eds) system dynamics. North-Holland, Amsterdam

König T, Jünemann K, Burger M (2015) RUnit-a unit test framework for R. Downloaded from https://cran.r-project.org/web/packages/RUnit/vignettes/RUnit.pdf. August 2015

Lo Giudice D (2013) Why agile development races ahead of traditional testing. Computer Weekly, 16–18. ISSN: 0010-4787

Peterson DW, Eberlein RL (1994) Reality check: a bridge between systems thinking and system dynamics. Syst Dyn Rev 10(2–3):159–174

Sterman JD (2000) Business dynamics: systems thinking and modeling for a complex world. Irwin/McGraw-Hill, Boston

Sterman JD (2002) All models are wrong: reflections on becoming a systems scientist. Syst Dyn Rev 18(4):501–531

Sücüllü C, Yücel G (2014) Behavior analysis and testing software (BATS). In: Proceedings of the 32nd international conference of the system dynamics society. Delft, The Netherlands

Van Vliet H (2008) Software engineering: principles and practice. Wiley, UK

Viboud C, Boëlle PY, Carrat F, Valleron AJ, Flahault A (2003) Prediction of the spread of influenza epidemics by the method of analogues. Am J Epidemiol 158(10):996–1006

Yücel G, Barlas B (2015) Pattern recognition for model testing, calibration, and behavior analysis. In: Rahmandad H, Oliva R, Osgood N (eds) Analytical methods for dynamic modelers. MIT Press, Cambridge

Chapter 7
Model Analysis and Calibration

The system dynamics approach leads to models with a large number of highly uncertain parameters, so we should ask ourselves which of the parameters are really important.

Andrew Ford and Hilary Flynn.
Statistical screening of system dynamics models.
System Dynamics Review. 21 (273–303).

Abstract This chapter introduces methods that support policy analysis for system dynamics models. First, a mathematical method for calculating loop polarity is presented, and this formal approach can be used to detect shifts in loop dominance, for example, when two feedback loops compete to influence a stock's value. Second, statistical screening is summarized, and this allows for an exploratory analysis of a system dynamics model in terms of analyzing which of the many uncertain parameters stand out as most influential. Third, model calibration is explored, which is a valuable technique based on optimization methods. This approach can be used to fit model parameters to historical data. In turn, this can improve client confidence, and also provide good parameter estimates that can form the basis of policy design and analysis.

Keywords Model analysis Sensitivity analysis Statistical screening Calibration

Model Analysis

As discussed earlier in Chap. 1, two important ideas underlying system dynamics are that: (1) The model represents a closed boundary around the system under study (Forrester 1968), and (2) the interaction of the model's structural elements (stocks, flows and feedback loops) are responsible for generating the system behavior (Sterman 2000). For example, in the SIR model, the stocks, flows and interaction of feedback structures provides a causal model and explanation for contagion dynamics. Understanding how these structural elements drive the model behavior is a challenging task, and the system dynamics research domain of *model analysis*

J. Duggan, *System Dynamics Modeling with R*, Lecture Notes in Social Networks,
DOI 10.1007/978-3-319-34043-2_7

provides a range of methods that can assist the policy design process (Duggan and Oliva 2013).

Richardson (1995), describes a mathematical approach for determining loop polarity, of which there are two types. Negative feedback generates balancing type behavior, where the direction of change for a stock is reversed due to the loop's influence. Positive feedback drives reinforcing behavior, as the value of a stock is amplified, and this generates exponential growth. For a one-stock system, the polarity of a feedback loop linking the inflow rate $\dot{x}$ and the stock x is shown in Eq. (7.1).

$$\textit{Loop polarity} = \text{sign}\left(\frac{d\dot{x}}{dx}\right) \tag{7.1}$$

This loop polarity equation represents the sign the derivative, where the stock x is on the x-axis, and the flow $d\dot{x}$ is on the y-axis. When these x–y values are plotted, this relationship is known as a *phase plot*, and provides important insights into how a system behaves.

This information is used to determine the loop polarity. A positive slope (sign = 1) indicates a positive feedback loop is dominant, whereas a negative slope (sign = −1) shows that a negative feedback loop is the dominant loop. The sign function is a useful transformation that converts any value into a set of discrete outputs, as shown in Eq. (7.2).

$$\text{sign}(x) = \begin{cases} -1, & x<0 \\ 0, & x=0 \\ 1, & x>0 \end{cases} \tag{7.2}$$

As a practical example, Eqs. (7.1) and (7.2) can be applied to a single stock feedback model of population growth, where the net flow is a constant (r) times the stock (x), as formulated in Eq. (7.3).

$$\frac{dx}{dt} = \dot{x} = rx \tag{7.3}$$

Taking the derivative of $\dot{x}$ with respect to x evaluates to r (7.4), which is the system's growth rate.

$$\frac{d\dot{x}}{dx} = r \tag{7.4}$$

Therefore the sign of the growth rate r (7.5) determines the polarity of the loop, as follows:

- If r is positive, it's a positive feedback loop, which drives exponential growth.
- If r is negative, this results in a balancing feedback loop, that leads to exponential decay.

The value r is also known as the *open loop gain* of the feedback structure. The gain refers to the strength of the signal returned by the loop, for example, a gain of 2 means that the change in a variable is doubled following each successive cycle through the feedback loop (Sterman 2000).

$$\text{sign}\left(\frac{d\dot{x}}{dx}\right) = \text{sign}(r) \tag{7.5}$$

The value of this approach is that it can identify when changes in dominant polarity occur in the simulation model, and so provides an insight into how the feedback structures influence system behavior. More formally, in summarizing the features of this analysis method, Richardson (1995) provides the following definition.

> In a first order system with level x and net rate of change $\dot{x}$, a shift in loop dominance is said to occur if and when $d\dot{x}/dx$ changes sign, that is, when the dominant polarity of the system changes.

Because the growth rate (r) in the population model is constant, there is no change in loop dominance for this initial example. However, the limits to growth model, presented in Chap. 3, and displayed again in Fig. 7.1, contains two competing feedback loops. One is reinforcing, which drives exponential growth, while the other is balancing, and provides a limiting factor to growth.

The R implementation of this model is summarized below. For this model, the initial stock is set at 100 (ensuring that growth can occur), the reference growth rate is 10 %, and the constraining capacity value is 10,000.

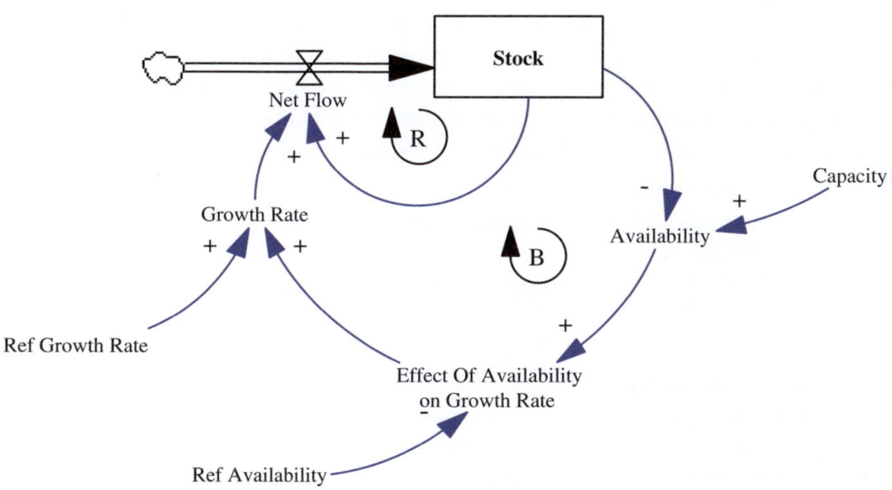

Fig. 7.1 Limits to growth, with two feedback loops

```
START<-0; FINISH<-100; STEP<-0.25
simtime <- seq(START, FINISH, by=STEP)
stocks  <- c(sStock=100)
auxs    <- c(aCapacity=10000, aRef.Availability=1,
                 aRef.GrowthRate=0.10)
```

The model function implementing the equations is now listed, with the growth rate is clearly influenced by the system availability. This ensures that the balancing feedback loop in represented in the model.

```
model <- function(time, stocks, auxs){
  with(as.list(c(stocks, auxs)),{
    aAvailability <- 1 - sStock / aCapacity
    aEffect       <- aAvailability / aRef.Availability
    aGrowth.Rate  <- aRef.GrowthRate * aEffect
    fNet.Flow     <- sStock * aGrowth.Rate

    dS_dt         <- fNet.Flow
    return (list(c(dS_dt), NetFlow=fNet.Flow,
                 GrowthRate=aGrowth.Rate,
                 Effect=aEffect,
                 Availability=aAvailability))
  })
}
```

To simplify the model analysis process, the loop polarity calculations for this model are performed numerically, based on the output from a simulation run. Two functions are specified to support the loop polarity calculation, starting with the function **deriv()**, which calculates a derivative, given a numerator and denominator. For this example, the numerator will be the net flow, and the denominator will be the stock. This function makes use of R's **diff(x)** function, which returns the differences between successive elements in a vector.

```
deriv<-function(yvar, xvar){
  ans<-vector(length=length(yvar))
  ans[1]<-NA
  ans[2:length(ans)]<-diff(yvar)/diff(xvar)
  return(ans)
}
```

The second function **polarity()** is used to determine the loop polarity, which will be either positive polarity ("POS") or negative polarity ("NEG"). This function accepts the net flow and the stock, calculates the derivative using **deriv()**, and determines the sign—using the R function **sign(x)**—according to the rules specified in Eq. (7.2).

```
polarity<-function(flow, stock){
  lsign<-sign(deriv(flow,stock))
  ifelse(lsign<0,"NEG","POS")
}
```

This limits to growth model is simulated with a call to the **deSolve** function **ode ()**, which populates an output data frame with the simulation results. The model analysis activity operates on these results. The net flow and stock values are then passed to the function **polarity()**, and the lop polarity classification is returned.

```
o<-data.frame(ode(y=stocks, times=simtime, func = model,
                  parms=auxs, method="euler"))

o$polarity<-polarity(o$NetFlow, o$sStock)
```

R's **ggplot()** function is then called, and the graph is colored (i.e. color = o$polarity) by the polarity attribute to effectively visualize the changing loop polarities over the different simulation intervals.

```
ggplot(o,aes(x=o$time,y=o$sStock))+
        geom_point(aes(color=o$polarity))+
        guides(color=guide_legend(title=NULL))+
        xlab("Time")+ylab("Stock")
```

Figure 7.2 illustrates the output, and shows how the polarity switches from positive to negative once the stock grows above 5000. In effect, this time point precisely captures the change in loop dominance, as the early positive feedback loop dominance is replaced by the limiting negative feedback loop. This point is

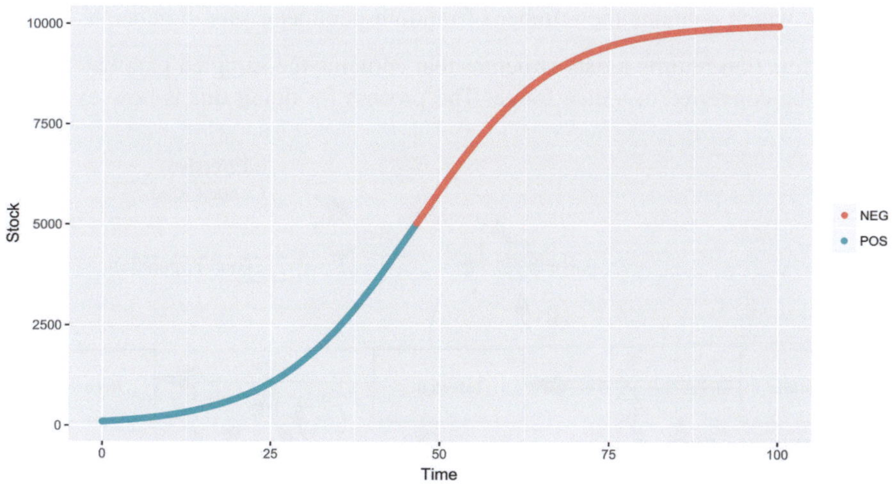

Fig. 7.2 Loop polarity analysis for the one-stock limits to growth model

commonly referred to as the point of inflection, and represent where the change in direction of curvature occurs.

Ricahrdson's (1984/94) loop polarity provides an excellent foundation for exploring addition model analysis methods, which are outside the scope of this text. These include: the *behavioral method* (Ford 1999), which identifies dominance by multiple loops and shadow loop structures; the *pathway participation metric* (Mojtahedzadeh et al. 2004), which shows which feedback loops are the most influential in explaining a selected pattern of behavior in a model; and *eigenvalue elasticity analysis* (Oliva 2015), which uses linear systems theory to decompose system behavior, and outline how the behaviors depend on system feedback loops. Another model analysis method, which does not use formal feedback loop analysis, is known as statistical screening, and makes use of available R functions for statistical and data analysis.

Statistical Screening

Ford and Flynn (2005) present a method to identify influential model parameters, through a process called statistical screening. The statistical screening process requires an initial sensitivity analysis, where a stock and flow model is run many times, with parameters sampled from a plausible range of values. An efficient method for sampling parameters is known as Latin Hypercube Sampling (LHS), which is effective for use in system dynamics modeling (Ford and McKay 1985). The R FME package (Soetaert and Petzoldt 2010) contains the function **Latinhyper(parRange, num)**, which takes two arguments, and generates a set of random parameter values.

- **parRange**, which is the range (min, max) for parameters. This contains a data frame with one row for each parameter, and two columns, one with the minimum value (1st column), and a second for the maximum value (2nd column).
- **num**, which contains the number of random parameter sets to generate.

This function returns a data structure that contains the sampled parameters, and this can be converted to a data frame. The process for doing this is now explored.

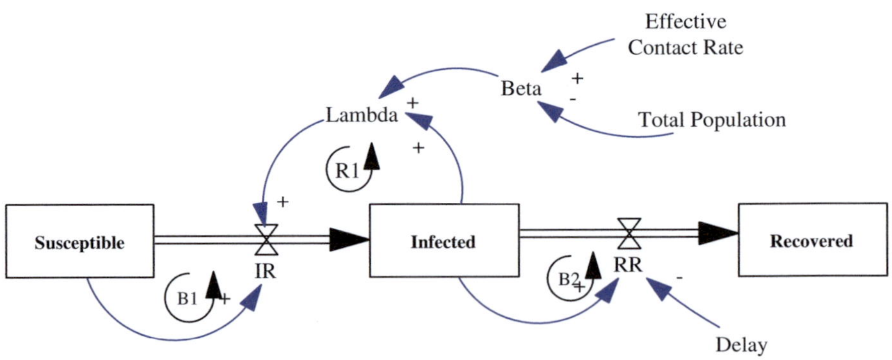

Fig. 7.3 The SIR model (equations specified in Chap. 5)

The aggregate SIR model—specified in Chap. 5—is used as an example. Its stock and flow structure is shown in Fig. 7.3, and the uncertain parameters include:

- The effective contact rate, C_E, which measures the level of contacts in the population, and the amount of contacts that lead to infection transmission.
- The recovery delay D, which models the amount of time it takes for individuals to recover from infection, where the recovery process is a first order delay.
- The initial number value of the infected stock.

For completeness, the R implementation of the SIR model is listed. This function will be called by the sensitivity analysis function, in order to create the required simulation data set for the statistical screening process.

```
model <- function(time, stocks, auxs){
  with(as.list(c(stocks, auxs)),{
    aBeta    <- aEffective.Contact.Rate / aTotalPopulation
    aLambda <- aBeta * sInfected

    fIR      <- sSusceptible * aLambda
    fRR      <- sInfected / aDelay

    dS_dt    <- -fIR
    dI_dt    <- fIR - fRR
    dR_dt    <- fRR

    return (list(c(dS_dt,dI_dt,dR_dt),
                 IR=fIR, RR=fRR,Beta=aBeta,Lambda=aLambda,
                 DEL=aDelay,CE=aEffective.Contact.Rate,
                 InitI=initInfected))
  })
}
```

The range of values for each parameter are defined, and these values would be selected in consultation with domain experts. For this example, an arbitrary range of values are identified.

```
CE.MIN<-0;                 CE.MAX<-7.0
DEL.MIN<-1.0;              DEL.MAX<-10.0
INIT.INF.MIN<-1.0;         INIT.INF.MAX<-25.0;
```

Based on these values, a data frame is created, which contains three rows and two columns, where each row refers to a parameter.

```
parRange<-data.frame(
  min=c(CE.MIN, DEL.MIN, INIT.INF.MIN),
  max=c(CE.MAX, DEL.MAX, INIT.INF.MAX)
)
```

Because each row in the data frame relates to a specific parameter, the row name is set to that parameter name.

```
rownames(parRange)<-c("aEffective.Contact.Rate","aDelay",
                      "initInfected")
```

The resulting data frame can be viewed, which clearly shows each parameter, along with its minimum and maximum value.

```
> parRange
                        min max
aEffective.Contact.Rate   0   7
aDelay                    1  10
initInfected              1  25
```

This data frame is processed by the **Latinhyper()** random number generator function. In this case just 5 random samples are generated.

```
> p<-data.frame(Latinhyper(parRange,5))
```

The resulting data frame contains the random numbers, all of which are LHS random variables that are within the specified ranges.

```
> p
  aEffective.Contact.Rate   aDelay initInfected
1               4.5385206 9.808586    16.813500
2               0.8180555 4.922799     9.968104
3               2.2954772 6.466728    13.240767
4               3.8191160 3.411032     5.329899
5               6.3927650 1.247923    21.493078
```

The next step is to write a sensitivity analysis function that takes, as input, this data frame, and returns a full set of simulation data for each random sample. First, a list structure is created that will store the simulation runs as a *list of data frames.* This is declared before the function is called, and the variable can be modified within the sensitivity function.

```
g.simRuns<-list()
```

The sensitivity function is named **sensRun(p)**, where the input value is the data frame populated with LHS parameter values.

```
sensRun<-function(p){

  g.simRuns<<-list(length=nrow(p))

  for(i in 1:nrow(p)){
    init     <-p[i,"initInfected"]

    auxs     <-c(aTotalPopulation=10000,p[i,1:3])

    stocks   <- c(sSusceptible=10000-init,sInfected=init,
                  sRecovered=0)

    o<-data.frame(ode(y=stocks, simtime, func = model,
                         parms=auxs, method="euler"))
    o$run           <-i
    g.simRuns[[i]]  <<-o
  }
}
```

The logic of the special purpose sensitivity function is as follows:

1. It creates a list storing all the simulation results, based on the number of rows in the input data frame **p**. A technical R point: the super-assignment operator **<<-** is used, as this sets the global variable value directly, and it is a useful way in R to avoid returning a large amount of data directly via the function return call.
2. It enters a loop, and iterates through each row of parameter data, with the variable **i** used as a loop index.
3. It extracts the appropriate parameter values from the data frame for the simulation run. Using matrix notation to access values, it starts with the variable **init**, which stores the initial number of infected in the population.
4. The vector **auxs** is then assigned, and contains the total population (a constant), and columns 1:3 of the row, which are the effective contact rate (aEffective. Contact.Rate), the recovery delay (aDelay), and the initial value for the infected stock (initInfected).
5. Following that, the vector containing stocks are declared and initialized.
6. The function **ode** is called, and the results stored in the data frame **o**.
7. The run number **(i)** is then added to the data frame **o**, as this is an important attribute to record, and is used when processing the simulation data at a later stage.
8. The full data frame is then added to the list **g.simRuns**, again using the super-assignment operator.

With the sensitivity function defined, just two commands are needed to run the ensemble of simulations. In this case, two hundred random samples of parameters are created using **Latinhyper()**, and then the sensitivity simulation in performed through a call to the function **sensRun()**.

```
p<-data.frame(Latinhyper(parRange,200))
sensRun(p)
```

When the sensitivity process is complete, the list *g.simRuns* contains all the results. However the list structure of 200 data frames is not convenient for overall simulation output analysis, and so a single data frame is created to store all the data. This is feasible, given that each simulation run has an identifier as a column. The R function **rbind.fill()** contained in the R library **plyr**, takes the list of data frames, and merges all these into one single data frame.

```
library(plyr)
df<-rbind.fill(g.simRuns)
```

The new data frame (**df**) can then be used to process the sensitivity results. For example, the following call to the **ggplot()** function groups the output by simulation run number, and so gives an immediate view as to the individual traces of the infected variable, across a run of 200 simulations.

```
p1<-ggplot(df,aes(x=time,y=sInfected,color=run,group=run)) +
  geom_line() +
  ylab("Infected") +
  xlab("Time (Days)")  + guides(color=FALSE)
```

The output in Fig. 7.4 is informative, as it provides an indication of the variation in simulation output, and shows where the behavior is mostly concentrated. However, the power of this sensitivity data set can be further harnessed through the use of the statistical screening method. Before describing the sequence of steps in R, the underlying theoretical framework for this valuable method is presented.

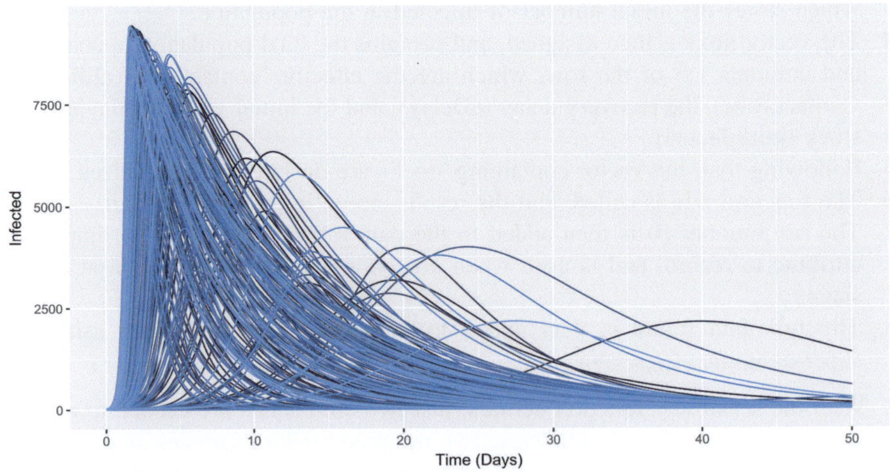

Fig. 7.4 Latin hypercube sampling applied to the SIR model

Statistical screening utilizes the sensitivity output data to calculate the *correlation coefficients* between parameters and a user-defined system performance variable (Taylor et al. 2010). This standard statistical measure (denoted r) determines the strength of the linear relationship between two variables (Groebner et al. 2011), and its formulation is shown in Eq. (7.6). The calculation is based on the time-series of two variables, X and Y. The correlation coefficient can range from a perfect negative correlation of −1.0, to a perfect positive correlation of +1.0. If two variables have no correlation, the value of r is zero.

$$r = \frac{\sum(X_i - \bar{X})(Y_i - \bar{Y})}{\sqrt{\sum(X_i - \bar{X})^2 \sum(Y_i - \bar{Y})^2}} \tag{7.6}$$

The statistic screening process calculates the correlation coefficient between two variables for each time unit of the simulation, and so provides a time series of values for each selected parameter against the variable of interest. The aim of this process is to identify the most influential parameters, and the following six steps are followed (Taylor et al. 2010):

1. Select a set of exogenous model parameters, and a system performance variable for analysis. Select appropriate ranges of exogenous parameters, based on an understanding of system being modeled.
2. Calculate the correlation coefficients between the selected exogenous model parameters and the system performance variables, using the statistical screening process. Plot the correlation coefficients and the behavior of the performance variable over time.
3. Select the time interval for analysis, by examining the time series data of both the performance variable, and the correlation coefficients.
4. Generate a list of high-leverage parameters, which are those that recorded the highest absolute correlation coefficient values during the selected time period.
5. Based on the parameters selected from step 4, identify the high-leverage model structure(s) that are directly influenced by the parameters. If additional parameters are connected to this model structure, then add each one to the list.
6. Develop explanations about how each parameter (or set of parameters), and the model structures they influence, drive the overall system behavior.

These steps are now followed for the SIR model.

Step 1: Select the exogenous parameters, and the variable of interest

For this example, the stock *Infected* is selected as the variable of interest, as it models disease prevalence in the population, and is important for epidemiologists and public health professionals. The exogenous parameters which influence this variable, already summarized from the SIR model, are highlighted in Table 7.1. This includes the initial value of the infected stock, the effective contact rate C_E and the average recovery delay D.

Table 7.1 Exogenous parameters for statistical screening with the SIR model

Parameter	Description	Min	Max
$Infected_{INIT}$	The initial value of number infected in the model. A number greater than zero is required in order for the disease to spread	1.0	25.0
C_E	Effective contact rate, where higher values increase the spread of a disease	0	7.0
D	Recovery delay, where a longer delay will result in people spending longer times in the infected stock	1.0	10.0

Step 2: Calculate correlation coefficients

The data frame **df** contains the sensitivity data needed to calculate the statistical correlation coefficients. However it requires additional processing, as the observations from each simulation time step need to be grouped together. This is achieved using R's **split(x, f)** function, which divides the data in the vector x into the groups defined by f. Therefore a list of 200 elements is created containing all the simulation data across the full set of runs *for each time step*. This command is shown below.

```
runs<-split(df,df$time)
```

The function **sapply()** is then used to process each list element, and calculate the correlation coefficient at each time interval. This function takes the list of data frames (organized by time step) as input. Its then takes each data frame, and calls the **cor()** function to calculate the correlation coefficient for the variable of interest, and the parameter. The following script calculates the correlation coefficient for the effective contact parameter.

```
cor.CE<-sapply(runs,function(l){cor(l$sInfected, l$CE)})
```

The returned value contains the correlation coefficient for these variables at each simulation time step. This is confirmed by exploring the vector *cor.CE*, where the length is the same as the simulation time vector, and the first 6 value are listed, rounded off to two decimal places with the **round()** function.

```
> length(cor.CE)
[1] 161
> round(head(cor.CE),2)
    0 0.125  0.25 0.375   0.5 0.625
 0.02  0.32  0.53  0.65  0.71  0.74
```

A similar procedure is followed for the additional model parameters.

```
cor.DEL<-sapply(runs,function(l){cor(l$sInfected, l$DEL )})
cor.initInf<-sapply(runs,function(l){cor(l$sInfected,
                           l$InitI )})
```

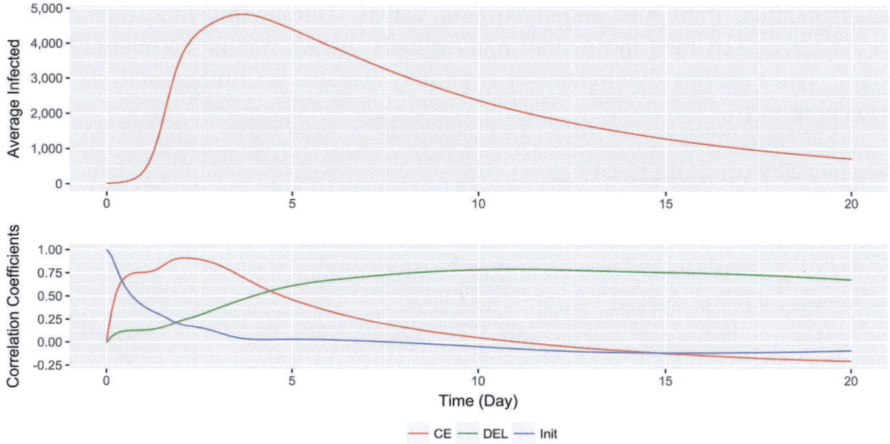

Fig. 7.5 Plotting the correlation coefficients and comparing with the variable of interest

The average value for the variable of interest at each time step is calculated, also using the **sapply()** function, and R's **mean()** function.

```
av.Infected<-sapply(runs,function(l){mean(l
$sInfected)})
```

A combined plot is created that shows how the different correlation coefficients vary over time, and their values can are aligned with the average behavior of the variable of interest. This provides a view on what the critical areas of the time horizon are in terms of model behavior, and supports the selection of the appropriate time interval (Fig. 7.5).

<u>Step 3: Select time interval for analysis</u>

Based on the simulation output, the appropriate time interval for the variable of interest is selected. For infection spread, the time of critical importance is the interval leading up to the peak value for the curve. In this case, when examining the average values across the 200 simulation runs, the interval [0, 5] captures the positive feedback driving exponential growth in the numbers of infected. In practice, the selection of time interval would also involve consultation with the clients, and the domain experts.

<u>Step 4: Generate list of high-leverage parameters</u>

During the selected time interval [0, 5], which accounts for the first 41 data points, a summary of each correlation coefficient can be obtained. This shows the mean, median, minimum and maximum values. During this time interval, the parameter C_E recorded the overall highest mean values for the correlation coefficient r, and therefore this parameter is initially selected for further analysis for the remaining steps.

```
> summary(cor.CE[1:41])
    Min.  1st Qu.   Median     Mean  3rd Qu.     Max.
-0.08106  0.61380  0.73430  0.69900  0.84250  0.91830

> summary(cor.DEL[1:41])
    Min.  1st Qu.   Median     Mean  3rd Qu.     Max.
-0.02631  0.06432  0.15220  0.19430  0.32750  0.49410

> summary(cor.initInf[1:41])
     Min.   1st Qu.    Median      Mean   3rd Qu.      Max.
-0.178400 -0.162900 -0.002348  0.110900  0.293900  1.000000
```

Step 5: Identify the model structure(s) influenced by the parameter

The next step is to revisit the stock and flow model, and observe how the parameter influences the structure. From Fig. 7.3, it is clear that the parameter C_E influences the contagion positive feedback loop structure. The information from the statistical screening process therefore supports the view that this is an important parameter.

Step 6: Develop explanations about how the parameter drives behavior

By exploring the equations and the feedback loop structure, the positive correlation between C_E and the variable of interest is confirmed. This parameter plays a key role in driving the exponential growth in the number of infected, as it directly increases the transmission parameter β, which in turn increases the force of infection value λ. Therefore, policy analysis could focus on measures that would dampen the influence of this parameter. Example could include actions for social distancing to reduce contacts between individuals, or indeed the use of personal protective equipment (PPE) which can reduce the probability of transmission when an infected person encounters a susceptible individual.

While it could be argued that the evidence from the statistical screening process for the SIR model, in terms of identifying the effective contact parameter, is self-evident, the value of the statistical screening method is that it is scalable. Higher order models with many parameters can be analysed, and the value of this process is that it provides a quantitative approach to identify the most influential parameters. Furthermore, utility functions in R such as **Latinhyper()**, **cor()** and **sapply()** can be deployed to generate the rapid analysis, and so provide useful insights as part of the model building process.

Model Calibration

In system dynamics, replicating system behavior using a stock and flow model is important, as it can increase user confidence in the model, and also assist with validation. The aim of model calibration is to fit the stock and flow model to past time series data (Dangerfield 2009). This involves exploring a parameter vector $\boldsymbol{p} = (p_1, p_2, ..., p_n)$ to determine the combination of values that provide the best fit between a designated model variable, and the historical time series of that variable. An optimization algorithm is used to explore the *search space* of the parameter set, in order to find the best fit. These fitted parameters then form the basis for validation and policy analysis.

The R algorithms used for calibration are based on the following functions from the FME package (Soetaert and Petzoldt 2010):

- **modCost()**, which estimates the residuals between model output and data. Here, for the given variables, the output from the simulation is compared to the time series data.
- **modFit()** which utilizes the output of **modCost()** to find the best-fit parameters, based on R's built-in optimization functions. The upper and lower bounds for the parameters are specified. Therefore, this function is used to find the optimal value that will find a best fit for the parameters, so that the model can replicate historical time series values.

This search process would be familiar to many system dynamics modelers. It is described by Coyle (1996) as *optimization through repeated simulation*. A schematic of the steps is shown in Fig. 7.6, where initial values of parameters are selected, and upper and lower bounds provided. The FME optimization function **modFit()** is called, and this organizes a search process to locate the best-fit parameters. The function **modCost()** calculates the accuracy of each solution by running the simulation with the parameter set, and evaluating the results against the available data. When the best set is found, **modFit()** terminates, and returns the best fit result $(PO_1, PO_2, ..., PO_N)$.

In order to demonstrate how the calibration process operates, a one-stock model of world population is used, and this is calibrated using historical time series data from 1960 to 2010. Shown in Fig. 7.7, the model has a single parameter (*growth*

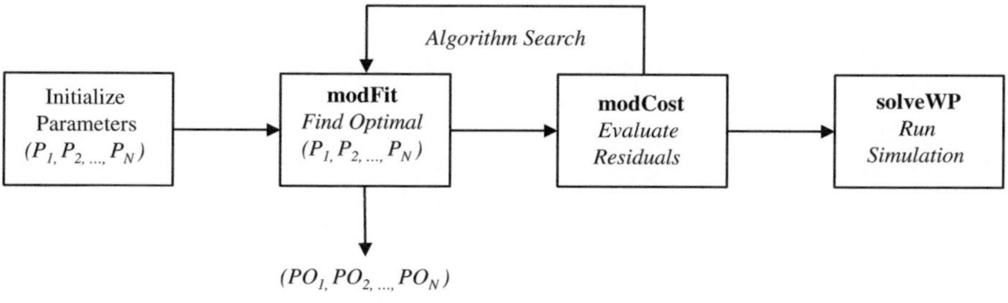

Fig. 7.6 Calibration process using R's FME libraries

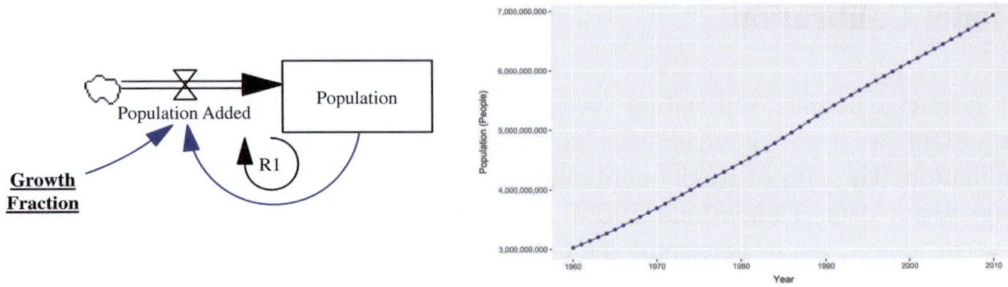

Fig. 7.7 World population model and historical time series data

fraction) that determines how fast the population grows, and the corresponding time series exhibits exponential growth properties, as the world population grew from about three billion in 1960, to over six billion in 2010.

The model equations are shown below. The stock (7.7) has one inflow, named population added (7.8), and the parameter to be estimated is the growth fraction (7.9).

$$Population = INTEGRAL(NetFlow, 3026002942) \tag{7.7}$$

$$Population\ Added = Population \times Growth\ Fraction \tag{7.8}$$

$$Growth\ Fraction = To\ be\ calibrated \tag{7.9}$$

The R script to calibrate this model requires a number of libraries:

- **deSolve** to run the simulation model.
- **gdata** to import the data from a spreadsheet using the function **read.xls()**, and in this example, the time series data is stored in the file *WorldPopulation.xlsx*. The sample code for reading the data is shown below, and is stored as a data frame. Sample output from the data frame is shown, where the two column names are *time* and *population*.
- **FME**, which provides the functions **modCost()** and **modFit()**.

```
library(deSolve)
library(gdata)
library(FME)

world_data <- read.xls("WorldPopulation.xlsx")

> str(world_data)
'data.frame':51 obs. of  2 variables:
 $ time : int  1960 1961 1962 1963 1964 1965...
 $ Population: num  3.03e+09 3.08e+09 3.14e+09 ...
```

This data frame *world_data* is important, as it will be used during the model calibration process. The simulation parameters are defined, which includes the start and finish times, and the simulation step. The initial stock value for population is also defined, which is the value specified in (7.7).

```
START<-1960; FINISH<-2010; STEP<-0.1;
simtime <- seq(START, FINISH, by=STEP)
WP_1960<-3026002942
```

The model function contains the necessary equations for running the simulation, which are implementations of (7.7) and (7.8).

```
model <- function(time, stocks, auxs){
  with(as.list(c(stocks, auxs)),{

    fPopulationAdded<-Population * aGrowthFraction

    dP_dt  <- fPopulationAdded

    return (list(c(dP_dt)))
  })
}
```

In order for the optimization process to operate, an additional R function is defined which accepts a parameter vector as its value. The function is named **solveWP()**, and it performs an important role in the optimization process.

```
solveWP <- function(pars){
  stocks  <- c(Population=WP_INIT)

  auxs    <- c(aGrowthFraction=
                   unname(pars["aGrowthFraction"]))

  return (data.frame(ode(y=stocks, simtime, func = model,
               parms=auxs, method="euler")))
}
```

The logic of **solveWP()** is as follows:

- It accepts a single vector with the list of parameters. This parameter list is provided through R's optimization process. In the case of the one-stock world population model, just one parameter, the growth fraction (7.9) is provided.
- It then initializes the stock vector, and sets the simulation parameter to the value contained in the vector **pars**.
- It calls deSolve's **ode()** function and returns the simulation data frame. This result is then returned by the function, and is then used to compare the simulation results against the actual values.

In order to run the calibration process, a final user-defined function, **getCost()**, is required which is used to compute the cost of an individual simulation run. In this context, cost refers to the fitness of a given parameter, and is based on the difference between the historical data values and the results returned by the simulation model.

```
getCost<-function(p)
{
  out<-solveWP(p)
  cost <- modCost(obs=world_data,model=out)
  return(cost)
}
```

The function **getCost()** performs three actions:

- It runs the simulation model with the parameter vector **p**.
- It calls the special purpose FME function **modCost()**, which computes the sum of squared residuals based on the simulation model output, and observed data.
- It then returns this cost value.

With these functions in place, the optimization process can now be completed. Three vectors needs to be declared, and these are **pars**, which contains the parameter names and their initial value, **lower**, which holds the lowest possible value for the parameter set, and **upper**, which provides an upper bound on each parameter value.

```
pars<-c(aGrowthFraction=0.001)

lower<-c(0.0)
upper<-c(0.2)
```

The function **modFit()** is called, and this accepts information on the parameters, and the target cost function.

```
Fit<-modFit(p=pars,f=getCost,lower=lower,upper=upper)
```

This function returns a list with the optimization result. Part of this list is the element **par**, which contains the optimized parameter value. In this case, it can be seen that the best fit growth fraction for the world population data from 1960 to 2010 is just over 1.75 %.

```
> optPar<-c(Fit$par)
> optPar
aGrowthFraction
     0.01754512
```

In order to enhance user confidence in the model, it is useful to plot both the actual data and the historical data on one plot. This can be completed by running an individual simulation run based on the optimal parameter value.

```
> optMod <- solveWP(optPar)
```

Following this, the simulation results can be filtered to select the model results for each year, by using the R **seq()** function to isolate the relevant row indices. This vector is then used to filter the simulation results.

```
time_points<-seq(from=1, to=length(simtime),by=1/STEP)
optMod<-optMod[time_points,]
```

The values are plotted using the following R code.

```
ggplot()+geom_point(data=world_data,size=1.5,
                    aes(time,Population,colour="Data"))+
         geom_line(data=optMod,size=1,
                    aes(x=time,y=Population,colour="Model"))+
         ylab("People")+
         xlab("Year")+
         scale_y_continuous(labels = comma)+
         theme(legend.position="bottom")+
         scale_colour_manual(name="",
                      values=c(Data="red", Model="blue"),
                      labels=c("Data","Model"))
```

A comparison between the fitted model and the historical data is shown in Fig. 7.8. The model simulates the historical value satisfactorily to the year 2000, but after that, the model value overestimates the time series value. This illustrates the strength and weakness of the calibration approach. On the one hand, it does provide a good estimate of what parameter value can drive the exponential growth, and this can improve user confidence in the model. However, it also highlights the need for a broader model boundary, as the exogenous variable (growth fraction) clearly has reduced over time. This would suggest that there are other factors at play in driving the growth rates, and models with more detailed stock, flow and feedback structures have been developed for this, including the formative world dynamics model by Forrester (1971), and the subsequent follow on model capturing limits to growth (Meadows et al. 2004).

Summary

This chapter introduced the idea of model analysis, which is a valuable part of system dynamics that can provide insight into how structural elements, namely stocks, flows and feedbacks, drive model behavior. Statistical screening is a practical model analysis method that can be used to identify which exogenous parameters have a significant influence on model behavior. R, through packages such as FME, and core correlation functions, supports the use of statistical screening, and this provides an excellent way to analyze large data sets containing simulation output. Furthermore, R also supports a calibration techniques to fit

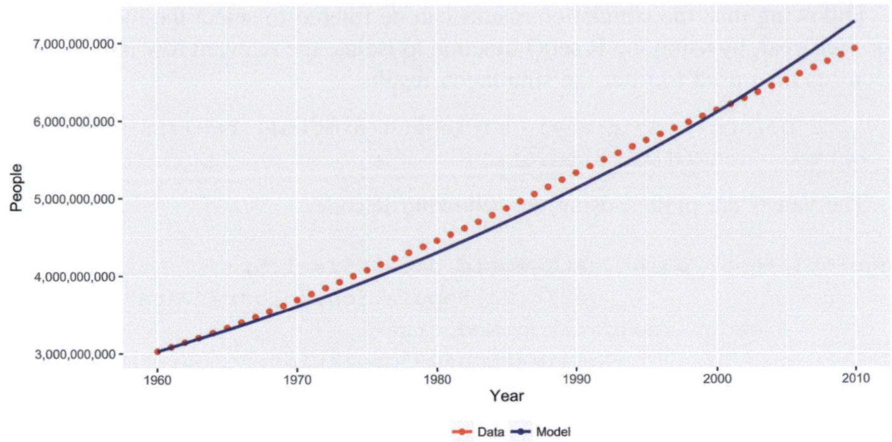

Fig. 7.8 Output from the calibration process

models to historical data, and this can enhance user confidence in models, and also provide good estimates to important model parameters.

Exercises

1. Calculate (formally) the loop polarity for the following system dynamics model, where r is the fractional decline rate, and T is the stock,

$$\frac{dT}{dt} = -rT$$

2. Use statistical screening to identify the most important parameters for the following customer model. Use the R model from this chapter as an exemplar for running multiple simulations, and for generating correlation coefficients. The simulation time runs from 2015 to 2035. Assume the growth fraction varies in the range [0.01–0.10] and the decline fraction's range is [0.01–0.08].

$$Customers = INTEGRAL(Recruits - Losses, 10,000)$$
$$Recruits = Customers \times Growth\ Fraction$$
$$Growth\ Fraction = 0.07$$
$$Losses = Customers \times Decline\ Fraction$$
$$Decline\ Fraction = 0.03$$

3. Consider the following empirical cooling data (Fahrenheit) for boiling water poured into a pot (Wagon and Portmann 2005), with time in seconds. Use R to calibrate this data to a suitable dynamic model. Assume the ambient temperature was 79°

Time	**0**	**25**	**50**	**75**	**100**	**125**	**150**	**175**	**200**
Temp	210	204	197	193	187	185	181	178	175
Time	**225**	**250**	**275**	**300**	**325**	**350**	**375**	**400**	**425**
Temp	172	170	168	166	163	160	159	157	155
Time	**450**	**475**	**500**	**525**	**550**	**575**	**600**	**625**	**650**
Temp	153	152	150	149	148	146	144	144	142
Time	**675**	**700**	**725**	**750**	**775**	**800**	**825**	**850**	**875**
Temp	141	140	139	138	137	136	135	133	133
Time	**900**	**925**	**950**	**975**	**1000**	**1025**	**1050**	**1075**	**1100**
Temp	132	131	130	129	128	128	127	126	125

References

Coyle RG (1996) System dynamics modelling: a practical approach. CRC Press, Boca Raton

Dangerfield B (2009) Optimization of system dynamics models. In: Meyers RA (ed) Encyclopedia of complexity and systems science. Springer, New York. ISBN 978-0-387-75888-6

Duggan J, Oliva R (2013) Methods for identifying structural dominance—introduction to the model analysis virtual issue. Syst Dyn Rev (Virtual Issue). http://onlinelibrary.wiley.com/journal/10.1002/(ISSN)1099-1727/homepage/VirtualIssuesPage.html

Ford DN (1999) A behavioral approach to feedback loop dominance analysis. Syst Dyn Rev 15(1): 3

Ford A, Flynn H (2005) Statistical screening of system dynamics models. Syst Dyn Rev 21(4): 273–303

Ford A, McKay MD (1985) Quantifying uncertainty in energy model forecasts. Energy Syst Policy (United States) 9(3)

Forrester JW (1968) Market growth as influenced by capital investment. Ind Manage Rev

Forrester JW (1971) World dynamics. Pegasus Communications, Waltham, MA

Groebner DF, Shannon PW, Fry PC, Smith KD (2011) Business statistics: a decision making approach. Prentice Hall/Pearson, Englewood Cliffs

Meadows D, Randers J, Meadows D (2004) Limits to growth: the 30-year update. Chelsea Green Publishing

Mojtahedzadeh M, Andersen DF, Richardson GP (2004) Using digest to implement the pathway participation method for detecting influential system structure. Syst Dyn Rev 20(1):1–20

Oliva R (2015) Eigenvalue elasticity analysis. In: Rahmandad H, Oliva R, Osgood N (eds) Analytical methods for dynamic modelers. MIT Press, Cambridge

Richardson GP (1995) Loop polarity, loop dominance, and the concept of dominant polarity (1984). Syst Dyn Rev 11(1):67–88

Soetaert KER, Petzoldt T (2010) Inverse modelling, sensitivity and monte carlo analysis in R using package FME. J Stat Softw 33

Sterman JD (2000) Business dynamics: systems thinking and modeling for a complex world. Boston: Irwin/McGraw-Hill

Taylor TR, Ford DN, Ford A (2010) Improving model understanding using statistical screening. Syst Dyn Rev 26(1):73–87

Wagon S, Portmann R (2005) How quickly does water cool. Math Educ Res 10(3)

Appendix A
Installing R and R Studio

R is a free software environment for statistical computing and graphics, and it compiles and runs on a wide variety of UNIX platforms, Windows and MacOS. To download R, access the web page https://www.r-project.org and follow the instructions.

Once R is installed, it is highly recommended to install R Studio.

R Studio is available on https://www.rstudio.com is an integrated development environment (IDE) for R, and is free and open-source. It provides an interactive workbench for creating, testing and running R scripts. This includes separate windows (see Fig. A.1) for:

- R scripts, containing the model equations and data processing scripts.
- An interactive console, for running, testing and debugging commands.
- The global environment, containing information on all variables stored in R's workspace.
- Access to the file system and graphical plots.

The R Studio IDE provides access to the following features.

- Easy installation of R contributed packages—such as the differential equation solver **deSolve**—through the menu option **Tools->Install Packages**.
- Full support for projects containing many source files (including a sub-directory structure) using the option **File->New Project**.
- Integration with code management systems such as **GitHub**, which is a powerful way to manage modeling projects, and also share system dynamics code with the wider community.
- Support for creating new R packages, which can include data and functions to support modeling efforts. For example, developers could export tested system dynamics models in package format, and these could be shared across the organization, and the wider community.

From a system dynamics modeling perspective, the following R packages, data structures and functions are most relevant.

J. Duggan, *System Dynamics Modeling with R*, Lecture Notes in Social Networks,
DOI 10.1007/978-3-319-34043-2

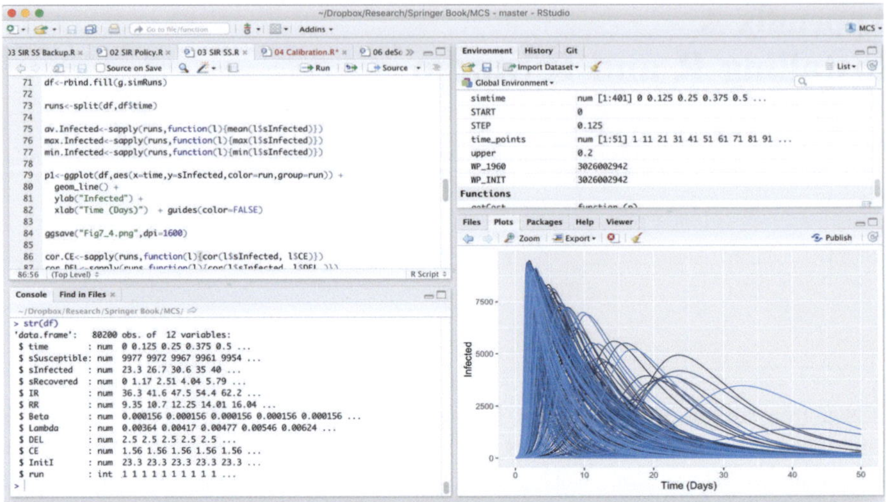

Fig. A.1 The R Studio IDE

- The package **deSolve** contains numerical integration functions, in particular, the function **ode()**.
- The package **FME** provides support for sensitivity analysis (latin hypercube sampling), and for model calibration, as illustrated through the examples in Chap. 7.
- The package **RUnit** provides a valuable testing framework for system dynamics models, as shown through the examples in Chap. 6.
- The package **ggplot2** provides an excellent way to visualize behavior over time, and generate high-quality charts that can enhance the model building process.
- The **data frame** stores simulation results and provides the key structure for analysis, as it is used by plotting functions.
- The **matrix** and associated operations supports vectorized models such as the SIR disaggregate model developed in Chap. 5.
- The function **cor()** calculates the correlation coefficient for two variables, and this function is required for as part of the statistical screening process.
- The function **approxfun()** supports the implementation of lookup tables in system dynamics models—see the example in Chap. 3.
- The function **sapply()** can be used to process all elements of a data structure, and return a vector as a result. This is used in Chap. 7, as part of the statistical screening process.

Glossary

Apply functions A family of functions in R that allows for efficient processing of vectors lists and data frames, by providing a function to operate on each data item. The most commonly used are: **apply()**, **sapply()**, **vapply()**, and **lapply()**

Atomic behavior pattern An essential shape of dynamic behavior based on the first and second derivative of a variable. It can be exponential, logarithmic, or linear

Auxiliary A model variable that is neither a stock or flow and usually a way to simplify flow equations, and make the model easier to understand

Closed system In system dynamics the idea that a model contains all the necessary stocks, flows and feedbacks to replicate the system's dynamic behavior

Data frame A data structure in R similar to a matrix, that can store different data types in each column

Delay Any process whereby the output lags the input. Delays always contain a stock and the order of a delay (i.e. the number of stocks) depends on the system being modeled

deSolve An open-source R package that solves initial value problems written as ordinary differential equations (ODE) differential algebraic equations (DAE), and partial differential equations (PDE)

Dimensional Analysis An equation-checking technique that ensures equations are balanced on each side of the equals sign

DT The time step of a simulation. As DT gets smaller the accuracy of the simulation improves

Effective contact rate In the SIR model this models contacts that are sufficient to lead to disease transmission, where that contact occurs between a susceptible person and an infectious person

Effects A system dynamics technique for modeling influences between two variables using an effect function

J. Duggan, *System Dynamics Modeling with R*, Lecture Notes in Social Networks,
DOI 10.1007/978-3-319-34043-2

Euler integration A process of numerical integration where the rates remain constant during the time interval DT

Feedback When circular causality is present in a system dynamics model this is known as feedback. Feedback represents a chain of connections from a stock, through flows, and back again to the original stock. There are two types of feedback. Positive feedback (reinforcing) amplifies change. Negative feedback (balancing) opposes change

Flow A flow is a rate that changes the value of a stock. All flows have the unit of time as their denominator. Lodgments/week is an inflow that causes a bank balance to rise

Force of infection In the SIR model this is the rate at which susceptible people become infected, per unit of time

Function An R programming construct that takes an input, processes the input, and returns an output value

ggplot2 A plotting library in R used to visualize behavior over time

Higher order model A system dynamics model with a significant number of stocks

Latin hypercube sampling A method to efficiently find representative parameters in a search space. Used to generate parameter ranges in sensitivity runs as part of statistical screening

Limits to growth The idea that a system cannot grow indefinitely and at some point a limit will be reached where further growth is not possible

Link polarity The relationship between a cause variable and an effect variable. A positive link polarity means that the two variables move in the same direction. A negative link polarity means that the variables move in opposite directions

List A data structure in R that can combine elements of different types

Loop polarity A calculation that determines whether a loop is a positive feedback loop or a negative feedback loop. If the number of negative links in a loop is odd, then the feedback loop is negative, otherwise the feedback loop is positive

Model building process An iterative five-stage process that provides a phased framework for constructing system dynamics models and is essential for implementing modeling projects with clients

Model calibration An optimization process that finds the best fit for model parameters in order that the model can replicate historical time-series values

Model validation The process of enhancing client confidence in system dynamics models. Can be classified into structural and behavioral tests

Mutation testing Changing a model equation from its original form in order to introduce an error. A useful way to test the efficacy of unit tests

ode() A special-purpose function is **deSolve** that performs numerical integration

Overshoot and collapse System behavior characterized by exponential growth followed by exponential decline as the resource base that fuels the growth is consumed, and not replaced

R Open-source software that has statistical data manipulation, and visualization libraries

R_0 The average number of secondary infections arising from one infectious person being added into a fully susceptible population

RUnit A package that supports unit testing for R programs

S-Shaped growth The classic growth behavior for a constrained system, characterized by exponential growth followed by logarithmic growth, as a system reaches its limit

Sector A sub-model with an overall system dynamics model that represents a coherent sub-system of the problem

SIR model A widely-used three stock model in epidemiology that models a virus as it spreads through a susceptible population. The infection rate is governed by the *force of infection* which depends on the number infected, and the effective contacts in the population

Solow Model A one-stock model of economic growth which captures the law of diminishing returns

Statistical screening A stepwise method to identify a model's most influential parameters. Requires sensitivity runs and then uses the correlation coefficient in order to identify the most influential parameters

Stock The building block of system dynamics models. A stock is an accumulation of some entity for example, money in a bank account, water in a reservoir. Stocks can only change through their flows

Stock management structure A stock and flow structure that models the regulation process for a stock and provides a formulation for the inflow (replacement) rate. This is based on an expectation of future losses, and an adjustment to move the stock towards its desired value

System dynamics A systems modeling methodology for building feedback models of social systems. The models may be qualitative or quantitative. Quantitative models are implemented using integral calculus and simulate the behavior over time of a social system

Vector A one-dimensional data structure in R that holds data of the same type

Index

J. Duggan, *System Dynamics Modeling with R*, Lecture Notes in Social Networks,
DOI 10.1007/978-3-319-34043-2

MIX
Papier aus verantwortungsvollen Quellen
Paper from responsible sources
FSC® C105338

If you have any concerns about our products,
you can contact us on
ProductSafety@springernature.com

In case Publisher is established outside the EU,
the EU authorized representative is:
Springer Nature Customer Service Center GmbH
Europaplatz 3, 69115 Heidelberg, Germany

Printed by Libri Plureos GmbH
in Hamburg, Germany